Hacking with Kali Linux

A Comprehensive Beginner's Guide to Learn Ethical Hacking. Practical Examples to Learn the Basics of Cybersecurity. Includes Penetration Testing with Kali Linux

ITC ACADEMY

Table of Contents

INTRODUCTION

Ethical Hacking - What Is the Need to Learn

This process is done by computer, and network experts called Ethical or white hat hackers. These people analyze and attack the security system of an organization to find and expose weaknesses that crackers can exploit and take advantage of it is essential to understand that unlike crackers, ethical hackers get permission from the relevant authority to go on and test the security of their information system. Cookies cause harm and loss to an organization and affect negatively to the integrity, availability, and confidentiality of an information system. So how did the concept of ethical hacking came up, and how is it done?

The field of ethical hacking has been there in the computer world now for a while. Today, this subject has continued to gain much attention due to the increased availability and usage of computer resources and the internet. This growth and expansion of computer infrastructure have provided another avenue for interaction, and so has attracted significant organizations in businesses and governments. These bodies want to fully take advantage of the benefits offered by technology so that they can improve the quality of service they provide to their customers. For example, organizations want to use the internet for electronic commerce and advertising, among others.

On the other hand, governments want to use these resources to distribute information to their citizens. Organizations fear the possibility of their computer information system being cracked and accessed by unauthorized people. On the other hand, potential customers and users of these services worry about the safety of the information they are prompted to give. They fear that this information, like credit card numbers, social security numbers, home addresses, and contacts, will be accessed by intruders or outsiders who are going to use their data for other purposes outside the ones that intended. By so doing, their privacy is going to be interfered with, something that is not desirable to many people, if not all.

Due to the above-raised fear, organizations sought to find a way to approach and counter this problem. They came to discover that one of the best methods they can use to limit and control the threat posed by unauthorized personal to a security system is to employ independent professionals in security matters to try the security measures of an order. In this scheme, hackers use the same tools and techniques used by intruders, but they do not damage the system, nor do they steal from it. They evaluate the system and report back to the owners the vulnerabilities their system is prone to. They also recommend what needs to be done to the system to make it more secure.

As evidenced from above, ethical hacking goes hand in hand with security strengthening. Though it has done much in increasing security matters, still more needs to be done. It is impossible to obtain absolute security, but even though, doing nothing to computer security is dangerous and undesirable.

CHAPTER 1

THE BASIC OF CYBERSECURITY

Cyber Security, Stay On Top Of The Silent Killer

Introduction:

Sometimes, we as humans tend to push toward advanced zones just for the sake of it, and in doing so, we usually miss out on some of the core areas and sticking to the basics.

To stay on top, it will be a solution from the top drawer from you. In tech-smart and highly developed markets, one as a business owner, operating online must ensure that things are under control all the time. These will save one from heavy penalties and consequential losses that usually take place in the form of data loss, compromise of business, and its clientele's sensitive information.

Such losses usually result in issues like stoppages and delays; hence, make it extremely hard for the business to cope with market competitions and client expectations.

Keep calm and stick to basics:

Although one will always have the facility of outsourcing such sensitive tasks to third-party IT security solution providers, it is still considered as a positive approach to getting the basics right from the start.

Experts associated with the domain of managed security services stress upon some of the very core areas that shall always be monitored by one as a business owner to ensure that things dealt with in a smooth and timely manner. Some of them are:

Passwords used by you must be strong:

Avoid setting secure passwords because they can be compromised easily. It becomes a cakewalk for smart hackers to figure out and crack such passwords before they get into your networks, systems, and online presence's code.

A secure password, therefore, is a must-have thing, and to ensure that you are on top of this requirement, you will need to make it a smart and robust blend of characters, alphabets, and numerals. By doing so, you are making it almost impossible for hackers to get into your system code.

Minimize the number of password attempts:

If you think that your six-digit password is enough to secure you, you must revise your approach. By ignoring this and opting for a six-digit pin, you are helping the attackers in creating more than a million unique possibilities to get into the sensitive domains of your business and damage the information. They have got hold of tools that will take just a few moments to crack such weak passwords.

You can rely on smart password managing software:

For some people, coming up with complicated passwords is a tough task. They merely believe that they are not good at it. They can use password managing software and obtain passwords with complicated combinations. These will help them in impressively securing the proceedings and information.

Prefer On-screen keyboard if working shared networks:

Hackers today have tools known as key-logging software; in shared network environments, a hacker uses this tool to record the keystrokes. To stay on top of this threat, one must prefer the on-screen keyboard while feeding in sensitive information.

Make backups regularly:

Experts associated with cybersecurity are of the view that one must ensure making regular backups. These will make it easy for you to restore the systems in case someone breaches in and tries to manipulate the data. Once done, one must not forget to change the passwords again. The activity of changing passwords must carry out regularly; relying on one password for a long time may not be classified as a smart approach.

Educate your staff about cybersecurity:

A well-trained team will make things easy for managed security solution providers. These will also help them to understand the instructions and do accordingly. One can save time, improve the processes, and potential cap threats for good. Business operations become smooth and secure if the staff members are trained.

Closing lines:

Technology is in its original form. Things are not going to stop here, though; they will continue to improve because this is an ongoing process. Hackers and attackers know this better than anyone else; therefore, they are always keen to stay on top. You can outsmart them by ensuring a regular backup in the form of managed services plus working in close collaboration with the experts that are hired by you.

Improving Your Cyber Security - A Beginner's Guide

A recent Eurobarometer poll revealed some relatively alarming statistics. Firstly, that around 10% of all European internet users had experienced online fraud or identity theft in some form, and that 74% of those quizzed believed cyber-crime to be 'increasing risk.' Secondly, that only just over 50% had some anti-virus software installed on their computers, and that 57% would open emails from addresses they did not recognize.

McAfee, in a separate study, has recently published a list of what it considers to be the foremost online threats in the coming year. Amongst the risks, it mentions employees of companies targeted as 'doorways' past security and more advanced viruses designed to steal banking information. These, coupled with the findings of the Eurobarometer poll, presents a worrying risk to European citizens. If those quizzed were aware of the inherent dangers posed by the internet, and yet did not take moves to protect themselves, then there are only a small number of possible reasons why. Firstly, they did not feel threatened by cyber-crime; however, given the 10% that had been victims of said crimes, it is unlikely that this is a universal principle. The second possibility is that there is a widespread lack of awareness when it comes to protection. The study found that even the most basic security protocols are ignored, so we shall, therefore, examine some security tips for the beginner.

Antivirus software - While your operating system of choice may have built-in software for dealing with specific threats, without a full, dedicated antivirus program installed, it can be hard to get frequently-updated protection against the ever-changing world of viruses. Many such programs can be bought cheaply, and charge an annual fee, but if you cannot afford them, then there are plenty of reliable providers with a free version of their product. Any protection is better than no protection.

Common sense - As a general rule-of-thumb, do not open emails that you either were not expecting or that are from addresses that you do not recognize. However, be cautious using email, regardless of the source. Some viruses can access your friends' email accounts, and use them to forward virus-riddled spam messages to everyone in that address' directory. If you receive a message out-of-the-blue from a friend, perhaps just quickly check that it is a genuine one.

Caution in security - Many websites and online services require users to register an account and provide a password, which is a good thing: they are protective of your details (and whatever else you keep there). However, be cautious about having one 'universal' password. If a hacker were to get hold of it, then they would have access to everything, and could even use it to change your details, and lock you out of your accounts. Repairing such damage would be very time-consuming.

Be wary of 'Cloud Storage' - Cloud Storage is a form of data-storage conducted solely online. It offers users either free or cheap use of its servers, and many people use it to 'file' documents and the like. It is an innovative and useful service. However, be wary of storing anything containing personal or confidential information in 'The Cloud', as it has gained a reputation for not being secure.

Learning Cyber Security Foundations

Computers have come to be an essential part of our life today. These require IT professionals, to have a good understanding of IT security foundations. These security foundations require an understanding of the controls needed to protect the confidentiality, integrity, and availability of the information.

Without healthy controls, cyber hackers and cybercriminals can threaten systems, expose information, and potentially halt operations. These types of attacks can create severe business losses. Cybercriminals and cyber hackers can target firewalls, IDS, and access control to enter the network and thereby causing severe damage. The problem of cybercrime gives rise to the need for cybersecurity training and dynamic controls to protect data. Anyone considering learning cyber foundations can learn the proper handling methods of sensitive corporate data.

The purpose of such training is to address aspects such as IT security and protection, responsibilities of people handling information, availability of data confidentiality, and how to handle problems such as unauthorized data modification, disruption, destruction, and misuse of information.

Any cybersecurity foundations training must cover aspects such as Network Security and Administration, Secure Software Development, Computer Forensics, and Penetration Testing. Here is a look at what such instruction should include as a minimum:

• Standard IT security terminology

• Current and future cybersecurity roles and positions that will be required by businesses to design secure IT computer networks successfully

• Aspects and concerns such as the hacker attack cycle and seven steps of cyber attacks

• The fundamental elements of security zones so that you are aware of defense in depth

• Auditing requirements.

With the internet bridging distances and making the whole world a global marketplace, computers have come to be only more important. Several companies offer online courses that cater to all the requirements related to computer security that an individual might have. Registering for a course that takes care of all your needs is essential. The course must be such that it equips you the power to take on the challenges of the present day competitive world. With knowledge by your side and the confidence to take on the planet, success is sure to be yours.

Cyber Security Made Easy

It seems like you can't watch the news without finding out about a new major security bug or corporate hacking scandal. Heartbleed and Shellshock scared many internet users, and soon articles on enhancing cybersecurity started popping up everywhere. Small business owners need to be especially savvy about cybersecurity since so much of their business base on the web. Here are some things you need to know about keeping your business safe online, as well as what to do in the event of a security breach.

• No business is too small to be vulnerable to hackers. According to the National Cyber Security Alliance, 71% of cyber attacks target small businesses, and almost half of the small businesses reported have attacked. Even more alarmingly, Experian has found that 60% of small businesses who are victims of a cyber attack go out of business within six months. The NCSA reported three reasons that small businesses are often targeted: they don't have the resources to respond to an attack, information like credit card numbers is often less heavily guarded, and small businesses may be partnered with larger corporations and give hackers access to those companies.

• Be sure that all devices dealing with the company network or any company data have reliable anti-virus and anti-malware software. These are basic, but easily overlooked, precaution against malicious files and other attacks. Your network should also have a firewall to protect the network as a whole.

• Educate your employees. In addition to making sure that everyone in your company is familiar with your security system, it may be useful to train employees on basic Internet safety and security. There are lots of online resources that raise awareness about phishing scams, security certificates, and other cybersecurity basics.

• Create strong passwords. For any resources requiring passwords on your system, create (and have employees create) complex passwords that aren't subject to social engineering or easy guessing. There are many guides available on the web about how to create strong passwords.

• Use encryption software if you deal with sensitive information regularly. That way, even if your data is compromised, the hacker won't be able to read it.

• Limit administrator privileges to your system. Set up the proper access boundaries for employees without administrator status, especially when using non-company devices. Limit administrator privileges to those who need them, and limit access to sensitive information by time and location.

• Look into cyber insurance. Cybersecurity breaches generally aren't covered by liability insurance, but if you're looking to protect sensitive data, talk to an insurance agent about your options.

• Back up your data weekly, either to a secure cloud location or to an external hard drive. That way, if your server goes down, you'll still have access to your data. Boardroom Executive Suites' Cloud Computing Services by SkySuite are an ideal tool in this area.

• If you've determined that there was a security breach, figure out the scope of the attack. These are an excellent time to call in a consultant who is an expert in cybersecurity. These will both give you a sense of what damage you need to mitigate and point to whether it was a generic mass-produced attack or a specifically targeted one.

• Once you've conducted this investigation, pull all of your systems offline to contain the damage.

• Repair affected systems. You can use master discs to reinstall programs on your devices. Then, with the help of your consultant, figure out where the gaps are in your security system. To prevent another attack from happening, use this as a learning experience to make your protection stronger.

• Be honest, transparent, and timely in your communication with your customers. Let them know what happened and what you're doing to fix it.

5 Cyber Security Mistakes Most Companies Make

Cybersecurity falls under the responsibility of everyone, not just information technology professionals. As with personal security, individuals must pay attention to their surroundings and their actions.

There are several areas that businesses and employees fail to pay attention to regarding cybersecurity. These are in no order of importance, as all are critical.

Lack of training for staff

When we raise our children, we make sure they know to look both ways before crossing the street, not to take candy from strangers, and never to get in a car with someone they don't know. To all of us, this is common sense as we received this same education ourselves.

With cybersecurity, the same principles apply. Don't open attachments from unknown sources. Don't go to websites that appear suspicious. Don't tell anyone your password(s).

Businesses must make sure they have education for all employees regarding these, and other, basic cybersecurity concepts. The training should occur at new hire orientation, and it makes sense to have annual or semi-annual reviews.

Failure to limit/log access

Who has access to what data? What IT Administrator modified the directory structure? Who changed permissions? Do all employees have access to HR files? Does any unnecessary person have access to financial records? Are there logs showing who accessed what data?

Most of the answers to these questions will be "we don't know," and that's a problem to acknowledge and address. Companies need to utilize built-in tools to log access, and, when necessary, purchase third party software for greater control and granularity. Not only can tracking access prevent a data breach, but it also enables organizations to find out what happened when data loss does occur.

Caring about corporate data

Most employees focus on their day to day jobs. They are not necessarily concerned with intellectual property at their company. Vast numbers of employees don't even know what data is critical to the success of their business.

Employees understand that financial and human resource records deserve protection; that's not enough.

Staff must also know about core data critical to the company so they can make sure and take proper action when dealing with that information and when dealing with others who have responsibility for protecting that data.

Understanding cyber threats

Phishing. Spoof. Worm. Trojan horse. Pharming. Hijack attack. All key terms in the cybersecurity world and, with few exceptions, most people do not know what these expressions mean.

Along with primary education, it makes sense for organizations to make sure the staff knows what these attacks are and how to protect against them. There are several terms and threats that individuals are familiar with; it's the responsibility of businesses to help employees understand additional dangers. Common sense goes a long way, and with adding simple communication, companies can ensure employees know what to look for and how to act when issues arise.

Spending money in the wrong areas, or not at all

Too often, businesses focus on revenue generation opportunities and ROI when spending money. Companies must take a defensive posture, as well. There doesn't mean only spending money on networking equipment and edge devices to protect their information assets, and they must understand the extent of the threats and spend in numerous areas.

Firewalls, extranets, and intrusion detection systems are all well and good; however, they only protect companies from specific types of attacks. Businesses must take a holistic view of cybersecurity and invest as necessary. Cybersecurity is an investment and should view as such through the budgeting process.

Everyone must take ownership of cybersecurity. In today's world, with significant data breaches occurring seemingly weekly, impacting millions of people, it's imperative to pay attention and share in the responsibility for data protection.

Through education, logging, corporate understanding data, knowledge of threats, and proper cybersecurity investments, companies will find greater security. When companies have data protection, investors, employees, and consumers receive peace of mind and clarity that they are as secure as possible.

CHAPTER 2

HOW TO INSTALL THE KALI LINUX AND HOW TO MAKE A KALI BOOTABLE USB DRIVE

At times, we have sensitive data; we would prefer to encrypt using full disk encryption. With the Kali Installer, you can initiate an LVM encrypted install on either Hard Disk or USB drives. The installation procedure is very similar to a "normal Kali Linux Install," except for choosing an Encrypted LVM partition during the installation process.

Kali Linux Encrypted Installation Requirements

Installing Kali Linux on your computer is a natural process. First, you'll need compatible computer hardware. The hardware requirements are minimal, as listed below, though the better device will naturally provide better performance. The i386 images have a default PAE kernel so that you can run them on systems with over 4GB of RAM. Download Kali Linux and either burn the ISO to DVD or prepare a USB stick with Kali Linux Live as the installation medium.

Installation Prerequisites

A minimum of 20 GB disk space for the Kali Linux install.

RAM for i386 and amd64 architectures, minimum: 1GB, Recommended: 2GB or more.

CD-DVD Drive / USB boot support

Preparing for the Installation

Download Kali Linux.

Burn The Kali Linux ISO to DVD or Image Kali Linux Live to USB.

Ensure that your computer is set to boot from CD / USB in your BIOS.

Kali Linux Installation Procedure

To start your installation, boot with your chosen installation medium. It would be best if you were greeted with the Kali Linux boot menu. Choose a Graphical or a Text-Mode install. In this example, we decided on a GUI install.

Select your preferred language, and then your country location. You'll also be prompted to configure your keyboard with the appropriate keymap.

The installer will copy the image to your hard disk, probe your network interfaces, and then prompt you to enter a hostname for your system. In the example below, we've entered "kali" as the hostname.

Enter a robust password for the root account.

Next, set your time zone.

The installer will now probe your disks and offer you four choices. For an Encrypted LVM install, choose the "Guided – use entire disk and set up encrypted LVM" option as shown below.

Choose the destination drive to install Kali. In this case, we chose a USB drive destination. We will use this USB drive to boot an encrypted instance of Kali.

Confirm your partitioning scheme and continue the installation.

Next, you will be ask for an encryption password. You will need to remember this password and use it each time to boot the encrypted instance of Kali Linux.

Configure network mirrors. Kali uses a central repository to distribute applications. You'll need to enter any appropriate proxy information as required.

NOTE! If you select "NO" on this screen, you will NOT be able to install packages from the Kali repositories.

Next, install GRUB.

Finally, click Continue to reboot into your new Kali installation. If you used a USB device as a destination drive, make sure you enable booting from USB devices in your BIOS. You will be ask for the encryption password you set earlier on every boot.

HOW TO MAKE A KALI BOOTABLE USB DRIVE

Making a Kali Bootable USB Drive

/02. Kali Linux Live /Making a Kali Bootable USB Drive

02. Kali Linux Live

Our fastest method and the most favorite method of getting up and running with Kali Linux is to run it "live" from a USB drive. This method has several advantages:

It's non-destructive — it makes no changes to the host system's hard drive or installed OS, and to go back to normal operations, you remove the "Kali Live" USB drive and restart the system.

It's portable — you can carry Kali Linux in your pocket and have it running in minutes on an available system

It's customizable — you can roll your custom Kali Linux ISO image and put it onto a USB drive using the same procedures

It's potentially persistent — with a bit of extra effort, you can configure your Kali Linux "live" USB drive to have persistent storage, so the data you collect is saved across reboots

To do this, we first need to create a bootable USB drive, which has been set up from an ISO image of Kali Linux.

What You'll Need

A verified copy of the appropriate ISO image of the latest Kali builds a model for the system you'll be running it on: see the details on downloading official Kali Linux images.

If you're running under Windows, you'll also need to download the Win32 Disk Imager utility. On Linux and OS X, you can use the dd command, which is pre-installed on those platforms.

A USB thumb drive, 4GB or more substantial. (Systems with a direct SD card slot can use an SD card with a similar capacity. The procedure is identical.)

Kali Linux Live USB Install Procedure

The specifics of this procedure will vary depending on whether you're doing it on a Windows, Linux, or OS X system.

Creating a Bootable Kali USB Drive on Windows

Plug your USB drive into an available USB port on your Windows PC, note which drives designator (e.g., "F:\") it uses once it mounts, and launch the Win32 Disk Imager software you downloaded.

Choose the Kali Linux ISO file to be imaged and verify that the USB drive to be overwritten is the correct one. Click the "Write" button.

Once the imaging is complete, safely eject the USB drive from the Windows machine. You can now use the USB device to boot into Kali Linux.

Creating a Bootable Kali USB Drive on Linux

Creating a bootable Kali Linux USB key in a Linux environment is comfortable. Once you've downloaded and verified your Kali ISO file, you can use the dd command to copy it over to your USB stick using the following procedure. Note that you'll need to be running as root, or to execute the dd command with sudo. The following example assumes a Linux Mint 17.1 desktop — depending on the distro you're using, a few specifics may vary slightly, but the general idea should be very similar.

WARNING: Although the process of imaging Kali Linux onto a USB drive is straightforward, you can just as easily overwrite a disk drive you didn't intend to with dd if you do not understand what you are doing, or if you specify an incorrect output path. Double-check what you're doing before you do it, it'll be too late afterward.

Consider yourself warned.

First, you'll need to identify the device path to use to write the image to your USB drive. Without the USB drive inserted into a port, execute the command

sudo fdisk -l

At a command prompt in a terminal window (if you don't use elevated privileges with fdisk, you won't get any output). You'll get output that will look something (not exactly) like this, showing a single drive — "/dev/sda" — containing three partitions (/dev/sda1, /dev/sda2, and /dev/sda5):

Now, plug your USB drive into an available USB port on your system, and run the same command, "sudo fdisk -l" a second time. Now, the output will look something (again, not exactly) like this, showing an additional device which wasn't there previously, in this example "/dev/sdb," a 16GB USB drive:

Proceed to (carefully!) image the Kali ISO file on the USB device. The example command below assumes that the ISO image you're writing is named "Kali-Linux-2017.1-amd64.iso" and is in your current working directory. The blocksize parameter can be increased, and while it may speed up the operation of the dd command, it can occasionally produce unbootable USB drives, depending on your system and a lot of different factors. The recommended value, "bs=512k", is conservative and reliable.

dd if=kali-Linux-2017.1-amd64.iso of=/dev/sdb bs=512k

Imaging the USB drive can take a good amount of time, over ten minutes or more is not unusual, as the sample output below shows. Be patient!

The dd command provides no feedback until it completed, but if your drive has an access indicator, you'll probably see it flickering from time to time. The time to dd the image across will depend on the speed of the system used, USB drive itself, and USB port it's inserted into it, and once dd has finished imaging the drive, it will output something that looks like this:

5823+1 records in

5823+1 records out

3053371392 bytes (3.1 GB) copied, 746.211 s, 4.1 MB/s

That's it! You can now boot into a Kali Live / Installer environment using the USB device.

Creating a Bootable Kali USB Drive on OS X

OS X is base on UNIX, so creating a bootable Kali Linux USB drive in an OS X environment is similar to doing it on Linux. Once you've downloaded and verified your chosen Kali ISO file, you use dd to copy it over to your USB stick.

WARNING: Although the process of imaging Kali on a USB drive is straightforward, you can just as easily overwrite a disk drive you didn't intend to with dd if you do not understand what you are doing, or if you specify an incorrect output path. Double-check what you're doing before you do it, it'll be too late afterward.

Consider yourself warned.

Without the USB drive plugged into the system, open a Terminal window, and type the command diskutil list at the command prompt.

You will get a list of the device paths (looking like /dev/rdisk0, /dev/rdisk1, etc.) of the disks mounted on your system, along with information on the partitions on each of the drives.

Plugin your USB device to your Apple computer's USB port and run the command diskutil list a second time. Your USB drive's path will most likely be the last one. In any case, it will be one which wasn't present before. In this example, you can see that there is now a /dev/disk6, which wasn't previously present.

Unmount the drive (for this example, the USB stick is /dev/disk6 — do not simply copy this, verify the correct path on your system!):

diskutil unmount /dev/disk6

Proceed to (carefully!) image the Kali ISO file on the USB device. The following command assumes that your USB drive is on the path /dev/disk6, and you're in the same directory with your Kali Linux ISO, which is named "Kali-Linux-2017.1-amd64.iso":

sudo dd if=kali-Linux-2017.1-amd64.iso of=/dev/disk6 bs=1m

Note: Increasing the block size (bs) will speed up the writing progress, but will also increase the chances of creating a bad USB stick. Using the given value on OS X has produced positive images consistently.

Imaging the USB drive can take a reasonable amount of time, over half an hour is not unusual, as the sample output below shows. Be patient!

The dd command provides no feedback until it's complete, but if your drive has an access indicator, you'll probably see it flickering from time to time. The time to dd the image across will depend on the speed of the system used, USB drive itself, and USB port it's inserted into it, and once dd has finished imaging the drive, it will output something that looks like this:

2911+1 records in

2911+1 records out

3053371392 bytes transferred in 2151.132182 secs (1419425 bytes/sec)

And that's it! You can now boot into a Kali Live / Installer environment using the USB device.

To boot from an alternate drive on an OS X system, bring up the boot menu by pressing the Option key immediately after powering on the device and select the drive you want to use.

CHAPTER 3

INTRODUCTION TO WIRELESS NETWORK HACKING AND HOW TO SCAN NETWORK

Wireless Network Type and Security

A wireless network connection is gaining popularity nowadays. This type of relationship is easy to set up, and most of the time, reliably provided that it set up correctly. Most of the houses and even offices are using a wireless network connection. However, an unsecured wireless network is dangerous from wireless connection stealers and tappers. Learn about two types of wireless network setup and two ways to secure your wireless connection.

Infrastructure wireless network connection

There are two types of wireless network connections. Infrastructure network connection and ad-hoc network connection. The infrastructure uses a router or a gateway that provides an IP address to each device and shares the internet connection. For example, you have a desktop and a laptop. For the computer or the desktop to connect wirelessly, the router or gateway should have wireless capability. The standard setup is the modem connected to the router, then the laptop and desktop could be connecting wired or wirelessly.

Ad-hoc wireless network connection

This type of relationship does not need any router or gateway. Ad-hoc connection involves two wireless devices that communicate directly. A good example would be two laptops. You can set up an ad-hoc connection on one of the computers. One of the machines will serve as a server and will broadcast its ad-hoc name. The second computer will scan for an ad-hoc signal and should detect computer number one. You will give an option to connect to the ad-hoc network.

Wireless Security

The wireless signal that is being broadcasted by your router can be secured or unsecured. The unprotected message allows an unauthorized user to connect to your sign and get a free internet connection. If the much-unauthorized user is connecting to your router, it will eat up your bandwidth resulting in slow internet. Not to mention, they can also hack into your computer since you already connected to one network.

Wired Equivalent Privacy (WEP), Wi-Fi Protected Access (WPA) & MAC Address

This type of wireless security is widely using due to its simplicity. Both WEP and WPA allow a user to produce a password, and from this password, it will convert to decimal or hexadecimal. The string generated from the password is longer. Thus they are secure.

Wireless adapter has a unique identification number called Media Access Control (MAC) address. Unregistered MAC addresses will not be allowed to connect by the router.

Secure Your Home Wireless Network And Keep Hackers Out

One of the most important, yet under-implemented features of any home wireless network is secure. Many people worry only about file sharing and network printing. But many people underestimate the need for security or do not understand it. Safety is important because unlike traditional wired networks, wireless signals are transmitted across a broader spectrum and thus can be easily picked up.

With the ever-growing threat of identity theft, you want to take the steps necessary to secure your home wireless network. There are some minor threats and significant threats to your personal information. On the small side, an intruder may easily connect to your unsecured wireless network and use it for free internet access. These are called piggybacking. These are minor threats to your privacy.

A significant threat is a hacker who is looking to attack your network and hack into it and access your files and personal data.

Once they have this, they can quickly start stealing your identity or selling it to someone who will.

Wireless security can quickly implement by following a few guidelines. One of the essential features is MAC address filtering. Each computer has a unique MAC address. In your network configuration, you enter the MAC addresses of your PCs and laptops. Any other equipment that specifies in your network will block you from accessing it.

Another vital network security feature is disabling the broadcast of the SSID, or name of your network. It is best to refer to the manufacturer's documentation for specifics. Most of the required security features built in the wireless router, and this is the first point that needs to be secured. Many of the security settings are turned off by default. Make sure you turn them on for the best security.

You can secure and manage your network yourself if you have the time, or you can buy a home networking software package that does it for you automatically. There are some excellent programs out there for under $30 that will secure and repair your home network.

Setting Up A Secure Home Wireless Network - 2 Simple Steps To Protect Your Privacy

Securing a home wireless network is a two-step process. The first step is to ensure network security by securing the wireless access point or router. All of this can usually do through the web-based software interface of the device by typing in the default address of '192.168.0.1'. or '192.168.1.1'.

The first thing that needs to be changed is the administrator's username and password. Many networks are hacked into, simply because nobody bothered changing the default values. These are the same as most VCRs still blinking noon because no one turned the time. Once it has done, enable MAC address filtering and add computers based on their MAC address. This option will allow only those specified computers to connect to the network but will not, however, guarantee total security.

The network SSID defines a name for the network. The default value of this should change to a loose and long string. This value should be written down in a safe place and entered into the machines that are allowed to connect. Disabling the broadcast of the network SSID also provides an extra layer of security as the network will not advertise itself to outsiders.

Encryption should also be enabled. The default encryption usually is weak and can break easily. Typically WEP (Wireless Equivalent Privacy) is used for data encryption. However, where available, try and use WPA-PSK encryption. This method uses

256-bit encryption for transmitting data, and the key also changes, so it provides a far more secure alternative to WEP.

Some routers have firewalls built-in. Where available, make sure that this option is enabled.

The second step of securing a home wireless network is securing each individual's PC. These could do by installing a software firewall (this may not always be necessary if a hardware firewall installed), antivirus software, anti-spam, and pop-up blocking software. It is essential to keep this software up to date, as downloading the latest security updates for the operating system and web browser you are using.

You can manage your home network yourself if you have the time, or you can buy a home networking software that will monitor your system and alert you to any intruders or weak security measures for about $30. Whichever way you choose to do it, make sure you secure your home wireless network.

Problems With Wireless Networks and Solutions for Cell Phones

We are seeing fewer wires now, thanks to the introduction of wireless networks. Cell phones have already replaced landlines to a great extent, and WI-FI is on the verge of overtaking wired networks.

There are many reasons why so many people are welcoming this service with open arms. Some advantages of wireless networks are:

• Give mobility to the device, which is, in fact, the most significant benefit.

• Cheaper than wire networks. One saves a lot of money online rent, wires, and installation.

• Easily accessible and reachable. Wires are hard to carry across rivers and mountains, whereas wireless networks face no such problem.

• They are flexible, and many devices can connect easily with not much labor required.

• Fast and easy for any temporary or permanent setup.

Now the other side to the coin. Wireless networks also have some disadvantages. Some of these disadvantages are:

• Lower speed at times when compared to a wired network.

• Less secure and easy to hack. Since the signals travel in frequencies, one can easily catch them using hacking techniques.

• Affected by surroundings, such as walls, fences, etc.

• Often, there are 'dead-zones' where no network coverage is available.

• Hugely affected by the weather. Weather conditions such as rain and storms affect it very negatively.

Companies are continuously working towards improving wireless network systems. Yes, there are disadvantages, but a lot of these cons can be removed by utilizing one of the various available solutions.

Some of the methods of doing away with the negatives and enjoying the wireless coverage are:

• Changing Network

Sometimes the culprit is the network provider. In such a situation, the only solution that people left with is to change the network. It is prescribed to confirm the quality of service that the provider provides so that such problems can be avoided as changing the network provider is often not an easy task.

• Checking Phone

Your phone might be the cause of distortion. Old phones with dysfunctional hardware and software can cause the network to perform at a subsidized level. It is always a good idea to keep your phone updated with the latest firmware and have it checked immediately in case of any issue.

• Installing Signal Boosters

No matter what device you are using, cell phone or a tablet, signal boosters are sure to work. Even if you are having your whole network on WI-FI and face poor signal reception, you can install a signal booster in your home and office. It dramatically enhances the ability of your device to catch signals, giving you a better experience.

Wireless networks are indeed better than wired networks in any case. In this scenario, the pros outweigh the cons by a considerable margin. The icing on the cake is the fact that these disadvantages can be easily removed by putting in a little effort.

Get yourself a wireless network service provider's service and, by utilizing the tips given above, enjoy full signal reception.

Wireless Network Solutions: Complete Coverage When You Need It The Most

As a prominent company head, you realize the importance of your employees being mobile, as they are required to attend numerous meetings and meet with clients out of the office. Today, where everything digitalized, it becomes imperative that your employees are equipped with wireless technologies. Hire wireless network expert services and ensure your employees are always connected.

For the success of your firm, your employees must create an excellent first impression. How embarrassing would it be if one of your efficient executives is on a lunch meeting with one of your prospective clients and when they switch on the laptop to demonstrate your company's success, they cannot even access the internet? Indeed, not making use of the excellent standards wireless network will earn your company negative points, and this could also lead to you losing out on the account.

In your bid to increase the mobility of your employees, you experiment with various hasty techniques for quick solutions. However, there are increased chances of them not working for you. An added problem with installing a flawless wireless network is that it requires the implementation of multiple procedures that complement one another, making it an extraordinarily time-consuming and somewhat complicated process. So how do you decide which rigorous solution will best suit your business and get you in the good books of your prospective clients?

An ideal solution is to rope in the services of an expert wireless network service, as they will assure that the WLAN is designed to become an integral and a rather logical component that contributes to your reliable network infrastructure. Also, ensure they offer the updated standard boost along with other fully secured wireless networks that are easy to measure and managed remotely. Thus ensuring you get a better return on your investment.

As you set out to look for a proficient service provider, make sure they kit your network with a highly compliant and robust wireless platform. These will help support your business's high-bandwidth applications, which will help provide your employees with greater mobility. Meanwhile, it also ensures they will be able to access high volume data at all times, thus enabling your employees to deliver enhanced presentations to your clients.

Also, to avoid high internet bills and to maintain privacy, it becomes essential that no one intercepts your Wi-Fi network. Nevertheless, attaining maximized security is not easy, especially with increasing incidences of hacking and data transferring by unauthorized sources. To control these illicit activities, it requires the implementation of rigorous and complicated security procedures, making it a complex and extremely time-consuming task.

A reliable wireless network solutions provider understands the need for a secured network aptly and so, makes sure they deploy the latest technologies that will maximize your network's security. Also, owing to their extensive experience, they can implement robust front line security measures for native TDM and Ethernet transmission on a single platform of your Point-to-Point and Point-to-Multipoint architectures flawlessly. They are thus assuring that only authorized individuals and devices can access your company network securely.

How To Protect Your Wireless Network Against Hackers

Creating a wireless network at your house is an excellent idea; it's straightforward to do, plus it enables you to surf the net conveniently from anywhere at your house. The problem is that lots of people are unaware of the danger that's added when using the home Wireless in an insecure manner.

If you want to protect your WiFi to prevent your neighbors from using up your bandwidth, then sure that's a sound cause to protect your WiFi. However, it mustn't be your primary worry. The most significant problem with insecure WiFis happens when a hacker is able to connect to your WiFi, if he succeeds on doing that then he can easily read the information that sent between you and the router and reveal your usernames, passwords or everything else that's sent between your devices, regardless of whether you are using SSL. This attack is known as "Man In The Middle" or MITM, and it can easily be performed by even a rookie hacker rather quickly. Securing your Wireless will significantly lessen the chance of this occurring. Almost all hackers that try to get into WiFi's will soon give up attempting to hack a secure wireless network as there are many insecure networks out there which provide a considerably simpler target.

All of the approaches that are going to be explained here require that you log in to your access point's user interface and change a few of its options. If you don't know how to do that, then go over to your access point manufacturer's web site and look for the guide for your particular model. Try to find information about how to gain access to that router's web interface.

1. Make use of a secure encryption

Using secure encryption is an essential course of action; this makes sure that you can only connect by using a password. Choosing these options is generally done from the security tab in the Wireless setup menu. You may typically pick from three or four options: Disabled, WEP, WPA, WPA2. You'll want to choose WPA2 or WPA - and absolutely under no circumstances choose disabled or WEP! WEP encryption is an extremely broken encryption algorithm that can be broken in a few minutes by a completely inexperienced hacker. If your router only supports WEP encryption but not WPA, then you should replace that router ASAP.

2. Utilize a strong encryption password

I'm sure you hear this phrase a lot "Use a strong password," and numerous people are thinking to themselves that on WiFis, it's not too crucial new flash - On WiFis, it's vital to use strong passwords. Its a piece of cake for a hacker to use a program that scans the WiFi for millions of password combinations in mere minutes and crack it. Use at least eight to ten characters and a mixture of numbers, special symbols, and letters. Avoid using dates, names, or phone numbers - This is the very first thing hackers try.

3. Alter the SSID name

The SSID is the name of your wireless network, and it's that name you choose from the list of nearby networks when you try to join. It's recommended that you alter that name and not use the default name. Modify it to a name that doesn't ultimately reveal who is the one who owns this wireless network, such as your surname - this is especially essential if you're living in a dense population area, for instance, an apartment building.

4. Restrict the Wifi's range

Restricting the transmit range will decrease the possibility of someone looking to hack your network. In many homes, the Wireless may be detected from outside of it, and there's generally no reason for this. Limit the transmission range, go outside your property, and look to find out if it could be discovered from there. When possible, you may relocate your router towards the center of the household to gain optimum coverage without creating blind spots caused by the restricted Wireless network range.

5. Alter the router's interface security password

The router's password is the password you need to type whenever logging in to the web interface. Technically if the hacker can connect to the access point's web interface then it's already too late, and the hacker has gained access to the wireless network already, but this is a good precaution step to take, and it will limit the control that the hacker has over your wireless network.

6. Check who is connected to the network

If you suspect that an intruder is using your WiFi, then most wireless routers have a feature that enables you to check the IP and MAC addresses of each device that's logged in to your access point. You can check out that list and make sure that you know each of the devices. This list is frequently named active DHCP list or something like that.

When you put into practice these pointers, then you've got a great chance versus even the most motivated hackers. Enabling these options takes little time, and the security gain is incredibly considerable. All of the tips are important. However, the most vital one is encryption password, in case you don't adopt this tip, then your network's security is completely broken.

How to Detect a Hacker Attack

Most computer vulnerabilities can be exploited in a variety of ways. Hacker attacks may use a single specific exploit, several exploits at the same time, a misconfiguration in one of the system components, or even a backdoor from an earlier attack.

Due to this, detecting hacker attacks is not an easy task, especially for an inexperienced user. This article gives a few basic guidelines to help you figure out either f your machine is under attack or if the security of your system has compromised. Keep in mind just like with viruses, and there is no 100% guarantee you will detect a hacker attack this way. However, there's a good chance that if your system were a hack, it would display one or more of the following behaviors.

Windows machines:

* Suspiciously high outgoing network traffic. If you are on a dial-up account or using ADSL and notice an unusually high volume of the sociable network (traffic, especially when your computer is idle or not necessarily uploading data), then your computer

may have compromised. Your computer may be being used either to send spam or by a network worm, which is replicating and sending copies of itself. For cable connections, this is less relevant - it is quite common to have the same amount of outgoing traffic as incoming traffic even if you are doing nothing more than browsing sites or downloading data from the Internet.

* Increased disk activity or suspicious looking files in the root directories of any drives. After hacking into a system, many hackers run a large scan for any new documents or files containing passwords or logins for bank or payment accounts such as PayPal. Similarly, some worms search the disk for files containing email addresses to use for propagation. If you notice significant disk activity even when the system is idle in conjunction with suspiciously named files in standard folders, this may be an indication of a system hack or malware infection.

* a Large number of packets which come from a single address being stop by a personal firewall. After locating a target (e.g., a company's IP range or a pool of home cable users), hackers usually run automated probing tools that try to use various exploits to break into the system. If you run a personal firewall (a fundamental element in protecting against hacker attacks) and notice an unusually high number of stopped packets coming from the same address, then this is a good indication that your machine is under attack. The good news is that if your firewall is

reporting these attacks, you are probably safe. In this case, the solution is to block the offending IP temporarily until the connection attempts to stop. Many personal firewalls and IDSs have such a feature built-in.

* Your resident antivirus suddenly starts reporting that backdoors or trojans have detected, even if you have not done anything out of the ordinary. Although hacker attacks can be sophisticated and innovative, many rely on known trojans or backdoors to gain full access to a compromised system. If the resident component of your antivirus is detecting and reporting such malware, this may be an indication that your order can be accessed from outside.

Unix machines:

* Suspiciously named files in the /tmp folder. Many exploits in the Unix world rely on creating temporary files in the /tmp standard folder, which is not always deleted after the system hack. The same is true for some worms known to infect Unix systems; they recompile themselves in the /tmp folder and use it as 'home.'

* Modified system binaries such as 'login', 'telnet', 'FTP', 'finger' or more complex daemons, 'sshd', 'ftpd' and the like. After breaking into a system, a hacker usually attempts to secure access by planting a backdoor in one of the daemons with direct access from the Internet, or by modifying standard system

utilities which are used to connect to other systems. The modified binaries are usually part of a rootkit and, generally, are 'stealth' against direct, simple inspection. In all cases, it is a good idea to maintain a database of checksums for every system utility and periodically verify them with the system offline, in single-user mode.

* Modified /etc/passwd, /etc/shadow, or other system files in the /etc folder. Sometimes hacker attacks may add a new user in /etc/passwd, which can be remotely log in a later date. Look for any suspicious usernames in the password file and monitor all additions, especially on a multi-user system.

* Suspicious services added to /etc/services. Opening a backdoor in a Unix system is sometimes a matter of adding two text lines. There is accomplished by modifying /etc/services as well as /etc/ined.conf. Closely monitor these two files for any additions, which may indicate a backdoor bound to an unused or suspicious port.

Securing Wireless Fidelities Against Infidels

Wireless internet has become the hippest form of going online. It is very convenient, and it affords its users extreme mobility as it is also compatible with mobile gears and gadgets. However, wireless networks are very vulnerable to intruder attacks, especially if you don't have it secured. With most American homes subscribing to wireless internet, the more important is it to obtain these networks and the computers connected to it, using different levels of security.

Level 1: Encryption

Encryption technologies scramble data and communications over the network, which makes your information unreadable to people who might try to hack into the system. These encryption technologies come in the form of passwords or passphrases that you have to key in before you are allowed to connect to a private network. WEP/WPA security ensures that only persons who know the passphrase will be able to connect to the system and take advantage of its internet connectivity.

When creating passwords for your network, avoid using your phone number, birth date, or anything that might contain easily guessed information about yourself. Creating strong passwords is the key to protecting your network from potential threats.

Level 2: SSID Broadcasting

By default, the default name of your network is your router's brand. This network name is also known as the SSID. Having them disabled will protect you from hackers, identity theft criminals, and other people with malicious motives who might hack into your network.

Disabling your SSID will make your network "invisible." No one will detect that your system is open, so if there are any hackers around scanning for policies that they want to victimize next, they will not scan yours.

Level 3: Pre-set password and username on router's configuration page

Each router page has its username and password. By default, the username is "admin," and the password is either "admin" or "password." There is one passcode that is universal to all router brands. Of course, any hacker or criminal will know these passcodes, so if they can connect to your network, they can tweak the configuration page of your router, and this can render your system or router, unusable.

Level 4: Firewall settings

Your router has its firewall that is built-in within the system. This firewall prevents other people from accessing your network and the computers connected to it. You can adjust your firewall settings to allow all applications in your network to get online, or adapt it accurately for your settings to allow only specific apps to connect. Your operating system, depending on its version and brand, may also have its firewall settings, which can also help secure your network.

Level 5: Anti-virus and anti-spyware software

It cannot be emphasized enough how important it is to protect your computer from viruses, spyware, and malware. Thousands of these malicious programs are created every day, and if you don't update your security software, chances are, you will still be prone to be infected because virus definitions are not updated on your computer.

How to Defend Your Computer Against Hackers

Whether or not you're using the internet all the time, it is still best to keep your data protected.

Here's a fact: Regardless if you just hooked up with an internet service provider last year, last month, or even just yesterday, you are still prone to getting malicious software into your system.

These forms of malicious software can get every data or activity that you do on your computer. Take a key logger, for example. Key loggers have their way of installing itself without you knowing it. The moment it's installed and runs in the background, it will capture every mouse and keyboard event that you do. So, if you happen to check your email - you enter your username and password - and your keylogger is active, expect

100% that it has already captured your email log-in details without you knowing it. Once the keylogger captures the information, it saves it as a text file, and again, without your knowledge, sends out your information to the hacker so he can log in to your email and check your inbox - and even send emails using your account!

Pretty scary? It is creepy. The good thing about it, though, there are ways to prevent this from happening. Most identity theft crimes, as we call it, are usually avoided by using the latest and most reliable internet security software. It is a minimal investment that you can use to protect your most precious data- your identity.

There are spyware removers, antivirus software, and internet security software that are sold online, which you can purchase and install into your computer. Like the malicious software that steals information, it also runs in the background to scan any illegal activity that's running in the background. The only thing you need to make sure of is to have it updated regularly. The updates are used to help detect any new malware or virus that has been developed. So, technically, if you scan your computer without installing the update first, it won't be able to detect in case a new malware has infected your system.

These forms of online protection can only do so much as far as protecting your personal information is concerned. With the different types of communication that are available online, it is still very possible for identity thieves and hackers to get what they want from you - especially if you only met them via chat, VOIP, or social networking sites. You still have the responsibility of filtering out the information that you give to strangers and online acquaintances. Never give out your personal information to people (or any entity) that you don't trust.

CHAPTER 4

KALI TOOLS

Kali Linux Penetration Testing Tools

Kali Linux contains a large amount of penetration testing tools from different niches of the security and forensics fields. This site aims to list them all and provide a quick reference to these tools. Also, the versions of the devices can track against their upstream sources. If you find any errors (typos, wrong URLs), please drop us an e-mail!

Penetration Testing Tools present in Kali Linux

Tools Listings

The Kali Linux penetration testing platform contains a vast array of tools and utilities, from information gathering to final reporting, that enable security and IT professionals to assess the safety of their systems.

Metapackages

Metapackages give you the flexibility to install specific subsets of tools based on your particular needs. Kali Linux includes metapackages for wireless, web applications, forensics, software-defined radio, and more.

Version Tracking

Maintaining and updating a large number of tools included in the Kali distribution is an on-going task. Our Version Tracking page allows you to compare the current upstream version with the version currently in Kali.

The Best 20 Hacking and Penetration Tools for Kali Linux

Hacking and Penetration Tools for Kali Linux

It is surprising how many people are interested in learning how to hack. Could it be because they usually have a Hollywood-based impression in their minds?

Anyway, thanks to the open-source community, we can list out many hacking tools to suit every one of your needs. Just remember to keep it ethical!

1. Aircrack-ng

Aircrack-ng is one of the best wireless password hack tools for WEP/WAP/WPA2 cracking utilized worldwide!

It works by taking packets of the network, analyses it via passwords recovered. It also possesses a console interface. In addition to this, Aircrack-ng also makes use of standard FMS (Fluhrer, Mantin, and Shamir) attacks along with a few optimizations such as the KoreK attacks and PTW attack to quicken the attack which is faster than the WEP.

If you find Aircrack-ng hard to use, check for tutorials available online.

Aircrack-ng Wifi Network Security

2. THC Hydra

THC Hydra uses a brute force attack to crack virtually any remote authentication service. It supports rapid dictionary attacks for 50+ protocols, including FTP, https, telnet, etc.

Hydra – Login Cracker

3. John the Ripper

John the Ripper is another popular cracking tool used in the penetration testing (and hacking) community. It was initially developed for Unix systems but has grown to be available on over 10 OS distros.

It features a customizable cracker, automatic password hash detection, brute force attack, and dictionary attack (among other cracking modes).

John The Ripper Password Cracker

4. Metasploit Framework

Metasploit Framework is an open-source framework with which security experts and teams verify vulnerabilities as well as run security assessments to better security awareness.

It features a plethora of tools with which you can create secure environments for vulnerability testing, and it works as a penetration testing system.

Metasploit Framework Penetration Testing Tool

5. Netcat

Netcat, usually abbreviated to NC, is a network utility with which you can use TCP/IP protocols to read and write data across network connections.

You can use it to create any connection as well as to explore and debug networks using tunneling mode, port-scanning, etc.

Netcat Network Analysis Tool

6. Nmap ("Network Mapper")

Network Mapper is a free and open-source utility tool used by system administrators to discover networks and audit their security.

It is swift in operation, well documented, features a GUI, supports data transfer, network inventory, etc.

Nmap Network Discovery and Security Auditing Tool

7. Nessus

Nessus is a remote scanning tool that you can use to check computers for security vulnerabilities. It does not actively block any weaknesses that your computers have, but it will be able to sniff them out by quickly running 1200+ vulnerability checks and throwing alerts when any security patches need to be made.

Nessus Vulnerability Scanner

8. WireShark

WireShark is an open-source packet analyzer that you can use free of charge. With it, you can see the activities on a network from a microscopic level coupled with pcap file access, customizable reports, advanced triggers, alerts, etc.

It is reportedly the world's most widely-used network protocol analyzer for Linux.

Wireshark Network Analyzer

9. Snort

Snort is a free and open-source NIDS with which you can detect security vulnerabilities on your computer.

With it, you can run traffic analysis, content searching/matching, packet logging on IP networks, and detect a variety of network attacks, among other features, all in real-time.

Snort Network Intrusion Prevention Tool

10. Kismet Wireless

Kismet Wireless is an intrusion detection system, network detector, and password sniffer. It works predominantly with Wi-Fi (IEEE 802.11) networks and can have its functionality extended using plugins.

Kismet Wireless Network Detector

11. Nikto

Nikto2 is a free and open-source web scanner for performing quick, comprehensive tests against items on the web. It does this by looking out for over 6500 potentially dangerous files, outdated program versions, vulnerable server configurations, and server-specif problems.

Nikto Web Server Scanner

12. Yersinia

Yersinia, named after the yersinia bacteria, is a network utility too designed to exploit vulnerable network protocols by pretending to be a secure network system analyzing and testing framework.

It features attacks for IEEE 802.1Q, Hot Standby Router Protocol (HSRP), Cisco Discovery Protocol (CDP), etc.

Yersinia Network Analyzing Tool

13. Burp Suite Scanner

Burp Suite Scanner is a professional integrated GUI platform for testing the security vulnerabilities of web applications.

It bundles all of it's testing and penetration tools into a Community (free) edition, and professional ($349 /user /year) edition.

Burp Security Vulnerability Scanner

14. Hashcat

Hashcat known in the security experts' community among the world's fastest and most advanced password cracker and recovery utility tool. It is open-source and features an in-kernel rule engine, 200+ Hash-types, a built-in benchmarking system, etc.

Hashcat Password Recovery Tool

15. Maltego

Maltego is propriety software but is widely used for open-source forensics and intelligence. It is a GUI link analysis utility tool that provides real-time data mining along with illustrated information sets using node-based graphs and multiple order connections.

Maltego Intelligence and Forensics Tool

16. BeEF (The Browser Exploitation Framework)

BeEF, as the name implies, is a penetration tool that focuses on browser vulnerabilities. With it, you can asses the security strength of a target environment using client-side attack vectors.

BeEF Browser Exploitation Framework

17. Fern Wifi Cracker

Fern Wifi Cracker is a Python-based GUI wireless security tool for auditing network vulnerabilities. With it, you can crack and recover WEP/WPA/WPS keys as well as several network-based attacks on Ethernet-based networks.

Fern Wifi Cracker

18. GNU MAC Changer

GNU MAC Changer is a network utility that facilitates a more comfortable and quicker manipulation of network interfaces' MAC addresses.

Gnu Mac Changer

19. Wifite2

Wifite2 is a free and open-source Python-based wireless network auditing utility tool designed to work correctly with pen-testing distros. It is a complete rewrite of Wifite and, thus, features improved performance.

It does an excellent job at decloaking and cracking hidden access points, cracking weak WEP passwords using a list of breaking techniques, etc.

Wifite Wireless Network Auditing Tool

20.Pixiewps

Pixiewps is a C-based brute-force offline utility tool for exploiting software implementations with little to no entropy. It was developed by Dominique Bongard in 2004 to use the "pixie-dust attack" to educate students.

Depending on the strength of the passwords you're trying to crack, Pixiewps can get the job done in a matter of seconds or minutes.

PixieWPS Brute Force Offline Tool

Well, ladies and gentlemen, we've come to the end of our long list of Penetration testing and Hacking Tools for Kali Linux.

CHAPTER 5

WHAT ARE VPN, TOR AND PROXY CHAINS AND HOW TO USE THEM FOR SECURITY

What is a VPN?

VPN stands for the virtual private network and is commonly used by organizations to provide remote access to a secure organizational system. For instance, you are working from home, and you need to access files in your computer at the office or connect to applications that are available only via your office network. If your office has VPN installed and your laptop or home computer is configured to connect to it, you can get what you need from the office without having to worry about the security of the data transported over the Internet.

VPNs are also used to mask the IP address of individual computers within the Internet. These allow people to surf the Internet anonymously or access location-restricted services such as Internet television.

Ordinary users would most likely be using a VPN in the second scenario. There are several VPN services offered over the Internet. For simple anonymous surfing, you can find a service as cheap as $5/month or even for free!

But to get the total anonymizing experience, a premium VPN account is advisable. Most providers make this as easy as possible for prospective subscribers - no IP numbers to configure into web applications, no software to install, easy to follow instructions on how to set up the VPN, etc.

That said, before subscribing to a VPN service, decide how you are going to use it: Is it merely for browsing web site content? Download torrents? Watch Internet television? Each provider has its terms and conditions for service, and some will include restrictions against "illegal" activities such as P2P file-sharing of intellectual property. It is best to look for and read the fine print before committing to anything.

Another tip is to look for providers who offer a trial period for prospective subscribers. These indicate that the provider has confidence in their product, and you will be able to judge if the service works for you.

Check the Internet speed: Is the rate reliable, or are there certain times of the day when browsing or downloading is inconvenient? Assess the timeliness and quality of their technical support: How long before they respond to questions or help requests? How well do they understand your concerns, and were they able to help? Bottom line: Is the service worth the cost?

Online forums are good sources of feedback on particular VPN service providers. Customer testimonials are well and good, but then you hardly read anything negative in those. With online forums, if you read much negative feedback on a provider from different people, that may be a sign that you shouldn't do business with that provider. In any case, you should try to get as much information as possible before subscribing.

Everything You Need To Know About VPN Services

What is a VPN? VPN is an abbreviation for the virtual private network. It can define as the method that is usually applied to add to the privacy and security into the public and private networks, the internet, and Wi-Fi hotspots.

VPNs are usually used by different kinds of corporations to enable them to protect any sensitive data that they may have. There has, however, been an increase in the use of the personal VPN option today. These can be attributed to the different transitions that are facing the internet today.

When you use a VPN, then privacy is improved to a considerable extent. The reason why you get better privacy with a BPN is the fact that the first IP address you may have been using is replaced with one that is provided by your VPN provider. There is an excellent way for subscribers to get an IP address from the gateway city that they may want, provided that the VPN provider offers it. You can use VPN to alter your location. You may be living in New York, but you can use a VPN to make it look like you are in London and so on. Each VPN provider offers different gateway cities that you can select from.

Security

The security features that are offered by VPNs are what attract most people. There are lots of methods that one can apply to intercept any data traveling to a given network. Firesheep and Wi-Fi spoofing are easy ways used to hacking any information that is needed. The analogy is the fact that the firewall will protect the data in the computer while the VPN will protect data even on the web.

Usually, the VPNs use highly advanced encryption protocols and the techniques that guarantee tunneling techniques that are secure to encapsulate different data transfers. Anyone who considers themselves as a savvy computer user may never use the internet without having a firewall as well as an antivirus that is updated.

Security is becoming very important to most people because security threats seem to be increasing. More and more people are also relying on the internet, which makes VPN even more attractive because they are the first round for purposes of security. Different integrity checks can be applied to make sure that data isn't lost and that the connection isn't hijacked in any way. All traffic is well protected, and the method is much preferred to the proxies.

The VPN setup

Setting up a VPN is a process that is quite straightforward. Usually, you only need a user name and the server address. Some smartphones are quite dominant, and they can configure the VPN using PPTP as well as L2TP/IPsec protocols. All the major OS can also set the PPTP VPN kind of connections. Getting a VPN may be the best idea that you may have for your business. Usually, the protocol numbers and the features that are offered grow as time passes. You may select the kind of VPN you need depending on what you require it for.

How VPNs Work

As more and more of us work on the move, from our homes or on personal devices that we bring into the office, it is becoming increasingly important to embrace the technologies behind the VPNs that allow us to 'remote on' to our office networks, giving us the freedom to take advantage of these flexible work practices. The following article provides a quick guide to how they work.

A Quick Definition

VPN is an abbreviation of Virtual Private Network and is a term that covers a whole range of technologies that allow users to securely connect to a network from a remote location via a public network, which, in practice, usually means the internet.

There are broadly two types of VPN. The first can be described as remote access and allows an individual user or device to access a network in another location across the internet. The second can be referred to as site-to-site and involves connecting a system in one place to a network in another.

The key feature of a VPN is that they allow communications between separate networks to be secure. That is, they allow data to travel between systems without being seen or accessed by those that should not be able to do so. To do this, a VPN needs to a) make sure the right people access the virtual network in the first place and b) prevent people from intercepting any data as it travels across the internet.

VPNs create what are termed 'tunnels' across the internet, through which the information can travel out of the reach of prying eyes, or sniffers as they are known. In the purest sense, tunnels involve the encryption of information at one end of the data transfer, and then it is decoding at the other.

They work by transferring encrypted packets of data across the internet and treating the sending and receiving computers as known devices (with predefined addresses) effectively on the same (albeit disconnected) network. To this end, the packets comprise an inner and outer package. The outer container has the job of transporting the inner pack across the internet from the gateway server on the sender's network to the gateway server on the receiver's web and, therefore, only contains information about the gateway servers to which it is going to and fro. If sniffers intercept the packets, they only see this information and not what the outer pack is transporting data, or which terminal computer/device it is heading for, as this is all encrypted. The encrypted inner packet contains the actual data that is being transferred and has further information on the address of the destination computer on the destination network as well as the sending computer on the sending network - both of which have been assigned IP addresses which define them as being on the same virtual (remotely connected) network. The outer packets are decrypted when they reach the VPN server on the destination network, and the inner packs are then routed to the correct destination computer.

It is analogous to putting a protective bubble around the encrypted inner packet while it is traveling across public networks - sniffers can see where this bubble is going, but they don't know what is in it. The balloon can only be peeled away when it reaches its destination network, while the specified destination computer can only decrypt its contents on that network.

There are a variety of technologies that are used to generate and interpret these encrypted packets such as IPSec (Internet Protocol Security) and TLS (Transport Layer Security) as well as a few proprietary technologies depending on the VPN provider in question. However, they all have the same purpose and aim: to allow computers to join remote networks across open public networks securely.

How Secure Is A VPN?

When it comes to Internet security, users should be careful. There are malicious parties at many Wi-Fi hotspots, waiting to hack into a user's personal information. Sensitive information such as emails, instant messages, and credit card information are all susceptible to being hacked if they are not adequately secured. These are where a Virtual Private Network, or VPN, comes in handy. However, most users who do not know what a VPN consists of are likely to question how secure they are. A VPN can protect users in ways that anti-virus software and firewalls cannot. Though these programs can use in conjunction with the VPN, they only protect the device itself. They do not protect data transmitted to or from the device.

Here are some advantages that a VPN can offer in terms of Internet security:

Prevent Deep Packet Inspection

Internet Service Providers, or ISPs, are known for tracking user activities online. ISPs use this information as a means of inspecting, throttling, and prioritizing the data that are sent to and from user devices. These mean slower speeds for users who don't connect with a VPN. When users access the Internet using a secure VPN, they defeat deep packet inspection. The ISP is not able to see the user's activities. They are only able to see that they are communicating with the VPN's server. All of the information the user sends and receives is encrypted and private.

Connect Via Wi-Fi without Worry

Users who connect to the Internet using mobile devices take advantage of free Wi-Fi connections at coffee shops, restaurants, airports, libraries, and more. Doing so puts the user's information at risk. Wi-Fi connections typically have no security. Anyone can use them. This means hackers and other malicious third parties can access the user's device with ease. They steal sensitive information and install malicious software of unprotected devices. Using a VPN can ensure that no matter where the user connects to the Internet, their data is not vulnerable.

Eliminate the Threat of Data Sniffers

A data sniffer is a software that can use both legitimately and illegitimately. Hackers use data sniffers to steal a user's personal information and other valuable information. These can include instant messenger conversations, sensitive credit card information, and emails. With a VPN, a data sniffer will only see scrambled details initially. It will not be able to decipher it.

Use on Various Devices

A VPN can use on many devices, including desktop computers, laptop computers, tablets, smart phones, and even many wireless routers. When setting up the VPN, there are protocol options that the user can choose from. The protocol used determines the speed, stability, and security of the VPN connection.

When a user connects with a VPN, the most crucial factor that is going to influence security is the VPN provider itself. If the provider offers all of these features and maintains its infrastructure, the user can be sure they are getting a capable and quality service.

VPN Safe Secure Services Protection

First of all, what is a VPN, and why do you need a VPN? You see, in today's world, security is a big concern. Thousands and even millions of dollars can lose because of a security breach. When you have top-secret or confidential information stolen, or if you have someone breaking into your computer system and deleting valuable information, you have a lot to lose.

So to prevent that from happening, innovative developers have come up with a solution known as VPN (Virtual Private Servers). A VPN is meant to protect your system, as well as your network from being compromised. In other words, you don't want anyone to have unauthorized access to your network. That will help keep your data safe as you work.

How a VPN works.

Here is a brief primer on how VPN works. When you surf the Internet, you are actually on a public network. That means that if your system is not protected whatsoever, anyone with malicious intent can create all sorts of problems for you. Some of these problems include spyware, viruses, intrusion attacks, etc.

To prevent that from happening, you can install additional hardware or software. When you install new equipment that acts as an active Firewall, that may help to stop many of the attacks. Some people prefer to do it with software, and that's where a VPN comes in.

VPN is short for Virtual Private Network. What happens is that you will be setting up a private tunnel that others will have difficulty locating. In other words, to the outside world, you appear as anonymous. You do this by connecting to another server, and this server acts as your connection to the public network. When you connect to the Internet this way, you are actually on a Private Network. This is a much more secure way to surf the Internet. And the solution is known as VPN.

Benefits of a VPN.

As you can imagine, there are many benefits for using a VPN. You enjoy a lot of security and privacy. For example, if you are surfing from a wireless hotspot, your user names and passwords may be sniffed by sniffers on the network. Sniffer software intercepts data transmitted over the system, and that's how your user names and passwords can capture. But when you connect through a VPN, there is no such risk.

Also, because you are surfing in a private network, you remain anonymous. Some websites log your IP address automatically, especially those that require form submissions. When you are in a closed system, your IP address cannot be tracked. This means that you get more privacy as you surf.

VPN Service Guide - Understanding the Importance of Having a Secure Internet Session

Whether you're at home, on the road, or in the office, access to a virtual private network is a beautiful thing to have. Keep those snooping ISP and government eyes off of your internet sessions. Even if you simply wish to be connected to as secure a network as possible when using Wi-Fi services so that you can conduct your financial transactions and other business tasks at peace, it's definitely worth investing in a VPN service.

As much as nobody likes to admit it, the internet itself is inherently insecure. There should always be extra precautions taken whenever necessary to ensure 100% security and protection when using the internet. Even routers themselves can be hacked or infected with some virus.

If you're unaware of how a VPN works, a simple way to put it would be that it enables the user to receive and send data while remaining anonymous and secure online. You can select a server from another part of the country - or even the world - and connect to it without also being physically present. If you are doing business in China, for instance, and want to connect to US sites that are banned by the Chinese government, a VPN service will allow you to do so.

Those involved in P2P sharing often use VPN networks so that they cannot be tracked. If you ever use torrent programs, then this is the best method for staying secure and anonymous. Virtual private networks are essential for businesses as well - mainly corporations and enterprises.

Should You Get a Free VPN Service?

What about a free VPN service? All the experts agree: "free" services should avoid. These are because the infrastructure to operate a system of virtual private networks is costly and must be paid for somehow. If the customer is not charge, then how is the provider getting the money to keep the system up and running? Probably by means we all hate, such as data gathering, advertisements, and other annoying reasons.

These don't mean you have to spend a fortune on a VPN plan, in any case. Some unusual ones keep the prices affordable. Usually, different methods are offered in the form of monthly or annual subscriptions.

You can usually expect to get much value out of a NordVPN subscription. It offers all the best features and the highest number of secure virtual servers around the world to choose from. Choose the right subscription for your needs and get a 30-day money-back guarantee.

VPN Networks and Security

On computer networks, information can be protected by encryption. Encryption means replacing the data with a scrambled string of nonsense. This nonsense can be turned back into the original data using the key shared by the two machines. This encryption is virtually unbreakable and, when it is used for business communications, it dramatically increases the level of safety that the business enjoys. It's also great for personal connections. VPN services use encryption, among other methods, to keep information safe.

Under the Radar

A VPN is frequently described as providing a way for users to create a secure tunnel over a public network. This analogy is pretty accurate in terms of explaining what's going on. The information exchanged over the VPN isn't visible to people on the Internet. This means that people on a VPN connection can reach their work resources, applications on a private computer, and many other types of information without having to worry about their data intercepted. There are plenty of uses for this technology, as you can imagine, but businesses are particularly heavy users.

Untraceable

The other form of security that VPN services provide is that of masking your IP address. Your IP address is the numerical address that servers use to send you the information you request. The VPN service routes you through a server that gives the websites you're visiting its IP address instead of yours. These prevent those websites from betting personal information from you and, of course, it makes it impossible for anyone snooping to say where you are.

Why This Matters for Security

There are plenty of ways that your IP address can be used against you. If someone with bad intentions knows that there's a business network set up at your IP address, they have a target. That target might be tested with a port scan, be the subject of DDoS attacks or have all kinds of other mayhem released upon it. Concealing your IP address is a significant way to protect your security online.

Having your data encrypted is also a big part of staying safe online. Until the computer revolution came around, everyday people could not get the type of security that's provided by modern encryption. Today, you can get encryption levels from VPN providers that make it nearly impossible for anyone to see your information.

If you're interested in upping your levels of security when you're surfing, consider adding a VPN service to the tools that you use. It's a powerful, meaningful, and effective way of increasing the level of security on your network, and, for your employees or for you, it's an easy way to access the information on your servers from anywhere in the world without exposing them to attack. These services are remarkably affordable and, if you need to obtain information from remote locations, it's a tremendous technological feature. Surfing for business or pleasure is much safer when you have control over your personal information and how it appears to others online.

Facts About VPN

VPN or virtual private network is typically used to provide employees remote access to a secure company network. An example would be an employee who needs to access the company's computer programs or applications or files that are only within the company's server.

If your company had a VPN connection installed in the company's system, and also in your owned laptop or desktop, you can get through your company's secure network and access all the files you need and acquire them without risking somebody else viewing the sensitive data. With a VPN connection, users or employees will have access to files, printers, and external drives located in the office without even going personally there.

Aside from the above uses, VPN can also mask the IP address of individual computers, making users surf the web anonymously, or access websites that are restricted only to one geographic location, such as TV online channels (NBC, ABC, and HULU) that can only be accessed within the United States. VPN finds a way around these restrictions, helping you be American anywhere in the world.

How Do I Get a VPN Connection?

Usually, there is no problem with the set-up process if the VPN connection is for company use. Owners hire professionals to do that kind of stuff. However, for personal use, you have to do the setting-up yourself.

Prices of VPN connection begins at $5 a month, such a small amount to pay for the protection that VPN can give you. You can choose from a long list of providers on the web. Once you sign up, you will be sent an email instructing you how to install the VPN on your computer.

Factors to Consider When Choosing a Provider

Before you choose a provider, know first what you will mainly use your VPN connection for. Would you be using it to access restricted channels? Are you going to use it for your small business where you have remote employees, or you need it for downloading torrents? You have to determine first your reason to match it with the right provider.

When choosing a provider, check if they have trial periods. That way, you will be able to 'test the waters before going in.' You can check if the speed is okay, if the privacy you need was provided, or whether the service/connection is reliable or not. When you feel satisfied after the trial period, you can sign up for a paid service, because you already know the type of service the VPN provider can give you.

Everything you wanted to know about Tor but was afraid to ask.

If you're interested in online privacy, then you've no doubt heard about Tor (The Onion Router). The Tor Network (or just "Tor") is an implementation of a program that was initially developed by the US Navy in the mid-1990s. It enables users greater anonymity online by encrypting internet traffic and passing it through a series of nodes.

Chances are, you have lots of questions about this project you'd like answered before you jump in. However, due to the negative associations, ' many people make with Tor and related projects, it's understandable that potential users are afraid to discuss their interests.

In this post, I'll ask (and answer) those questions for you. I'll explain everything you need to know about Tor, including how anonymous it is, whether it's legal, and if you still need to connect to a VPN while using Tor.

What is Tor, and how does it work?

The Tor network, often referred to as "Tor," is a volunteer-run system that helps make internet use more anonymous.

When a user is connected to Tor (often through the Tor browser), their outgoing internet traffic is rerouted through a random series of at least three nodes (called relays) before reaching its destination (the website the user wants to visit). Your computer is connected to an entry node, and the final node traffic passes through is the exit node, after which it reaches its destination (the website you want to visit). Incoming traffic is rerouted similarly.

A simplified version of how Tor works (Source: EFF via Wikimedia)

Aside from passing through several nodes, the traffic is encrypted, multiple times in fact. It loses a level of encryption at each node but is never fully decrypted until it leaves the exit node for its destination.

Each node has an identifying IP address, which is also encrypted. The only IP address visible to the destination website is that of the final node, known as the exit node.

In total, the Tor network is currently made up of about 7,000 relays (nodes) and 800 bridges. Bridges are similar to relays, but they are not listed in the Tor directory. They may also be used as a website or app blocks traffic from a detected Tor node.

Does Tor hide IP address?

While connected to the Tor network, the activity will never be traceable back to your IP address. Similarly, your Internet Service Provider (ISP) won't be able to view information about the contents of your traffic, including which website you're visiting.

Your ISP will see that you're connecting to a Tor entry node, and the website you're visiting will know the IP address of the Tor exit node.

How to use Tor: getting started

The simplest way to use Tor is through the Tor browser. These are a Firefox-based application that can be downloaded and installed on your computer.

Versions are available for macOS, Windows, and Linux. Once you've downloaded and installed, you'll be able to access clearnet and .onion sites through the browser.

In some cases, the use of the Tor browser may block. As mentioned earlier, using a bridge should overcome this issue. In the past, this was fairly complex but is a lot easier in the latest version. You'll need first to locate a bridge and then configure it with the Tor browser.

Does Tor make you anonymous?

It's complicated, if not impossible, to become truly anonymous online, but Tor can undoubtedly help you get there. All of your traffic arriving at its destination will appear to come from a Tor exit node so that it will have the IP address of that node assigned to it. Because the traffic has passed through several additional nodes while encrypted, it can't be traced back to you.

However, one of the issues lies in trusting the operator of the exit node. If you're visiting an unencrypted (non-HTTPS) website, it's possible the node operator can track your activity and view your information. They could collect data such as which webpages you're considering, your login information, the content of your messages or posts, and the searches you perform. Although, there's no way to trace that information back to you or even back to the entry node.

It's worth noting that using the Tor browser only protects traffic going through that connection and won't anonymize other apps on your computer (although many can configure to the Tor network via other means). Also, your ISP can still see that you're using Tor. For improved privacy, you can use a VPN alongside the Tor browser.

What is the darknet, and how is Tor related to it?

If you're familiar with the term, the "clear net," you'll know that it refers to the portion of the internet that can be freely accessed, that is, without Tor or an alternate browser. On the other side, you have the deep web. These include content that isn't indexed by search engines, including outdated content, private files, and web pages that have barred search engines from crawling them.

Also, within the deep web is the darknet. This content can usually only be accessed using special tools like Tor. The darknet houses some legitimate websites, but it is better known for being a place rife with illicit activity.

You can access the clear net with Tor, but you can also access darknet websites, specifically .onion sites. These are sites that only people using the Tor browser can access and have .onion as part of their URL. They are also referred to as "Tor hidden services."

They're not indexed by search engines and can be challenging to find if you don't know where to look. Tor protects the anonymity of the operators of .onion sites, so it would be difficult to find out who is running them. Of course, the combination of both operator and user anonymity is what makes the darknet ideal for criminal activity.

The (now seized) website for the infamous AlphaBay marketplace was a .onion site. (Source: US Department of Justice via Wikipedia)

That being said, there are plenty of legitimate websites that have .onion versions. For example, VPNs are geared toward privacy-conscious users, and some offer .onion versions of their site, ExpressVPN being one example. You can even set up a .onion site of your own through the Tor browser.

Why would someone want to use Tor?

As mentioned, Tor is often associated with illegal activity and users wanting to access the dark web. Because of this, there is often an assumption that anyone using Tor must be up to no good. On the contrary, Tor can be used by privacy-conscious users for day-to-day browsing on clearnet sites, to help maintain user anonymity and privacy while online.

There are a vast number of reasons your average internet user might want to be more anonymous. These include stopping ISPs and third parties collecting data about online activity, bypassing censorship, protecting children's privacy, or researching taboo topics, such as birth control or religion.

There are also many professions in which it would be necessary or helpful to keep an anonymous online profile. Some of those legitimately using Tor include:

Journalists

Law enforcement officers

Activists

Whistleblowers

Business executives

Bloggers

Militaries

IT professionals

Although Tor doesn't track what users are doing online, it does offer aggregate statistics telling you where users are located. You can see graphs by country and read about events that may have contributed to drastic changes in user numbers.

For example, the above graph shows the number of US users connecting over the past year. In the dated commentary below each chart, Tor provides notes about things like updates, outages, and significant events such as government blockages.

Is using Tor legal?

The nature of Tor indeed makes it a popular choice among criminals wanting to access some of the shadier parts of the darknet and conduct criminal activities. These include buying or selling illegal products or services or participating in forums that spread hate speech and encourage extremism.

However, as outlined above, there are plenty of reasons non-criminals would want to use Tor. Indeed, it is perfectly legal to use Tor, although it has been or is blocked in certain countries. Plus, there is still a stigma attached to it, so you probably shouldn't assume you can use it trouble-free.

ISPs have been reported to throttle the bandwidth of Tor users and have even contacted customers to tell them to stop using the Tor browser. Users may question by ISPs regarding which websites they are connecting to through Tor.

Authorities themselves could become suspicious of Tor users and conduct investigations into their activities on those grounds alone, although there haven't been reports of fines or charges related to the use of Tor.

Are there any downsides to using Tor?

Tor is popular with many users — there are currently around 2 million users connecting to relays at a given time.

But it does have its downsides. Here are the main cons of using Tor:

Slow speeds

Detectable by ISPs

Blocked by network administrators

Vulnerable to attacks

Let's look at each of these in a bit more detail.

Slow speeds

The major downside to using Tor is that it is quiet. Traffic isn't going directly to its destination, so this will slow things down. Plus, the speed of traffic flowing between the nodes could be slower than your regular internet connection, further dampening the overall momentum.

What's more, the number of volunteer nodes available is minimal compared to the amount of traffic flowing through the network. The resulting congestion will slow down traffic, especially during peak periods.

Due to these issues, the primary use of Tor is general browsing. It isn't suitable for streaming or torrenting, or anything else that requires much bandwidth.

Detectable by ISPs

Another downside is that your ISP will be able to see that you're using Tor. It won't be able to read the contents of your traffic, but the fact that it detects you're using Tor could have some repercussions. As mentioned earlier, using Tor alone is enough to raise suspicion from ISPs and authorities. One way around this is to use a VPN with Tor (more on that below).

Blocked by network administrators

Administrators of individual networks often block tor. One way around this is to use bridges that shouldn't be detectable as Tor nodes. If the blockage is more sophisticated and uses deep packet inspection, you may need to use an additional tool, such as Pluggable Transports (see below). These will mask your Tor traffic as regular traffic to bypass the block.

Vulnerable to attacks

While it hasn't been confirmed, there have been reports that traffic analysis on Tor has been successfully used to find incriminating evidence. One case that stands out is the Silk Road takedown of 2013. Silk Road was a marketplace run through the Tor network and was involved in the sale of an estimated $1 billion worth of drugs, along with other illicit goods and services.

There are also the rumors mentioned above about exit node monitoring to be wary of. Bear in mind that these reports don't appear to have been confirmed so that they can be viewed with skepticism.

How to protect your privacy online with Tor Browser

If you want to keep your web browsing private, you can use the Incognito mode in Chrome, Private Browsing in Firefox, InPrivate mode in Microsoft Edge, and so on. While this will prevent other people who use your computer from seeing your browsing history, it doesn't restrict your ISP from monitoring the sites you are visiting. You might well want to – for any number of reasons – browse the internet completely anonymously, and this is precisely what Tor Browser offers.

Standing for The Onion Router, Tor offers multiple levels of protection to ensure that your online activities, location, and identity are kept entirely private. Here are the steps you need to follow to install and use Tor Browser.

1. Install and configure Tor Browser

Start by downloading and installing the Tor Browser. Click Finish once the installation is complete, and Tor will launch for the first time. You'll be greeted by a settings dialog that is used to control how you connect to the Tor network.

In most cases, you should be able to click the Connect button, but if you connect to the internet through a proxy, you will need to click the Configure button to enter your settings.

2. Get online with Tor

There will be a slight delay while Tor establishes a connection to the network via relays – the program warns that the initial link could take as long as several minutes – but once this connection has been made, the Tor browser will launch ready for use.

Tor is based on the same code as Firefox, so if you have used Mozilla's web browser, everything should seem reasonably familiar. Even if you haven't used Firefox before, it should not take you long before you start to feel at home – it's not so different from the likes of Edge, Chrome, and Safari.

3. Choose your security level

Before you get started, it's worth noting that using Tor Browser is a balancing act between privacy/security and web usability. By default, security is set to Standard – although this is still far more secure than any other web browser.

If you would like to increase this, click the onion icon to the left of the address bar and select Security Settings. Use the Security Level slider to choose your preferred level of protection, bearing in mind the warnings that appear about the features that may stop working on the sites you visit.

4. Rethink your browsing habits

To get the most from Tor, you need to change a few of your browsing habits – the first of these is the search engine you use.

Rather than opting for the likes of Google and Bing, the recommendation is that you instead turn to Disconnect.me. This is a site that prevents search engines from tracking you online, and you can use it in conjunction with Bing, Yahoo, or DuckDuckGo.

While we're on the subject of changing habits, you also need to avoid installing browser extensions, as these can leak private information.

5. Understand Tor circuits

As you browse the internet, the Tor browser helps to keep you secure by avoiding direct connecting to websites. Instead, your connection is bounced around between multiple nodes on the Tor network, with each jump featuring anonymizing.

This not only makes it all but impossible for a website to track who and where you are, it is also responsible for the slightly slow performance you will notice while browsing with Tor.

If you feel performance is unusually low or a page is no longer responding, you can start a new Tor circuit by clicking the hamburger icon and selecting the 'New Tor Circuit for this Site' option, which will force Tor to find a new route to the site.

6. Create a new identity

The new circuit option only applies to the currently active tab, and it may be that you want a more drastic privacy safety net. Click the hamburger icon and select 'New Identity,' bearing in mind that this will close and restart Tor to obtain a new IP address.

When you connect to a site using Tor, you may notice that a pop-up appears warning you that a particular site is trying to do something that could potentially be used to track you. Just how often these messages appear will depend not only on the websites you visit but also on the privacy settings you have in place.

7. Use HTTPS

An essential part of staying safe and anonymous online is ensuring that you use the HTTPS rather than HTTP versions of websites. So you don't have to remember to do this for every site you visit, Tor Browser comes with the HTTPS Everywhere extension installed by default. This will try to redirect you to the secure version of any website if it is available, but you should keep an eye on the address bar as an extra safeguard.

If you are connected to a secure site, you will see a green padlock icon. If this isn't present, click the 'i' icon for more information.

8. Access .onion sites

The most secure way to connect to the internet through Tor, however, is to visit .onion sites. These are also known as hidden Tor services, and they are inaccessible to search engines; to find them, you have to visit them directly.

These sites can only be accessed using Tor, but you do need to take care − it's quite common to come across sites with illegal content, selling illegal products, or promoting illegal activities.

9. Try Tor over VPN

If you want to take your privacy to the next level, you can connect to a VPN before starting the Tor browser. The VPN will not be able to see what you're doing in the Tor browser, and you'll get the added benefit that no Tor node will be able to see your IP address. It will also prevent your network operators from even knowing that you are using Tor, which is helpful if the Tor Network happens to be blocked where you are.

How to Use Proxychains to Redirect Traffic Through Proxy Server

Sometimes we install a proxy server, but only specific programs such as Firefox and Google chrome provide proxy settings. Luckily, we can use a command-line utility called proxy chains to redirect any application to go through our proxy server. This tutorial will show you how to set it up on Debian, Ubuntu, OpenSUSE, Fedora, CentOS/Redhat, Arch Linux, and their derivatives.

If you don't know how to set up a proxy server, then check out this post to learn shadowsocks proxy. After that, come back here.

Install proxy chains on Linux

Debian/Ubuntu/Linux Mint/Elementary OS

sudo apt-get install proxy chains

OpenSUSE Leap 42.1

Proxy chains are available from the packman repository.

sudo zypper install proxy chains

Fedora

sudo dnf install proxy chains

CentOS/Redhat

sudo yum install proxy chains

Archlinux

sudo Pacman -S proxy chains-ng

On Kali Linux, proxy chains are installed by default.

Add a Proxy Server to Proxychains

Open the configuration file.

sudo nano /etc/proxychains.conf

At the end of the file, add your proxy like this

socks5 127.0.0.1 1080

socks5 is the proxy type, and you can add other types as well, such as HTTP, https, socks4, etc. depending on your situation.

127.0.0.1 is the proxy host, and 1080 is the port on which the proxy server listens. Again, change them to your specific situation.

The default proxy is socks4 127.0.0.1 9050, which you can safely remove.

Set a Default DNS Server

It's highly recommended that you change the default 4.4.2.2 DNS server to something else, such as Google's DNS server 8.8.8.8/8.8.4.4. Or OpenDNS server 203.67.222.222/203.67.220.220. Open the resolv configuration file.

Debian/Ubuntu

sudo nano /usr/lib/proxyresolv

Linux Mint/Elementary OS

sudo nano /usr/lib/proxychains3/proxyresolv

Fedora/CentOS/Redhat/OpenSUSE

sudo nano /usr/bin/proxyresolv

Find the following line

DNS_SERVER=4.4.2.2

Change its value to something like 8.8.8.8. Then save and close the file. On Arch Linux, there's no proxyresolv config file.

Test

Just prepend proxychains to any command you execute like the following.

proxychains youtube-dl -citw
https://www.youtube.com/channel/<channel-id>

If you are using youtube-dl, then you may know that it has no built-in support for socks proxy, but Proxychains will redirect youtube-dl to go through the proxy server.

Quiet Mode

By default, proxychains will output its activity to the terminal. If you don't want to see this information, then you can disable it by editing /etc/proxychains.conf file.

CHAPTER 6

HOW TO HACK A WIRELESS NETWORK

How to Hack WiFi (Wireless) Network

Wireless networks are accessible to anyone within the router's transmission radius. These make them vulnerable to attacks. Hotspots are available in public places such as airports, restaurants, parks, etc.

In this tutorial, we will introduce you to conventional techniques used to exploit weaknesses in wireless network security implementations. We will also look at some of the countermeasures you can put in place to protect against such attacks.

What is a wireless network?

A wireless network is a network that uses radio waves to link computers and other devices together. The implementation is done at Layer 1 (physical layer) of the OSI model.

How to access a wireless network?

You will need a wireless network-enabled device such as a laptop, tablet, smartphones, etc. You will also need to be within the transmission radius of a wireless network access point. Most devices (if the wireless network option is turned on) will provide you with a list of available networks. If the system is not password protected, then you have to click on connect. If it is password protected, then you will need the password to gain access.

Wireless Network Authentication

Since the network is easily accessible to everyone with a wireless network-enabled device, most systems are password protected. Let's look at some of the most commonly used authentication techniques.

WEP

WEP is the acronym for Wired Equivalent Privacy. It develops for IEEE 802.11 WLAN standards. Its goal was to provide the privacy equivalent to that provided by wired networks. WEP works by encrypting the data been transmitted over the network to keep it safe from eavesdropping.

WEP Authentication

Open System Authentication (OSA) – this method grants access to station authentication requested based on the configured access policy.

Shared Key Authentication (SKA) – This method sends to an encrypted challenge to the station requesting access. The station encrypts the problem with its key then responds. If the encrypted challenge matches the AP value, then access is granted.

WEP Weakness

WEP has significant design flaws and vulnerabilities.

The integrity of the packets is checked using Cyclic Redundancy Check (CRC32). A CRC32 integrity check can compromise by capturing at least two packs. The bits in the encrypted stream and the checksum can be modified by the attacker so that the authentication system accepts the packet. These lead to unauthorized access to the network.

WEP uses the RC4 encryption algorithm to create stream ciphers. The stream cipher input is made up of an initial value (IV) and a secret key. The length of the initial value (IV) is 24 bits long, while the secret key can either be 40 bits or 104 bits long. The total length of both the initial value and secret can either be 64 bits or 128 bits long. The lower possible value of the secret key makes it easy to crack it.

Weak Initial values combinations do not encrypt sufficiently. These make them vulnerable to attacks.

WEP is based on passwords; this makes it vulnerable to dictionary attacks.

Keys management are poorly implement. Changing keys, especially on large networks, is challenging. WEP does not provide a centralized key management system.

The initial values can be reused

Because of these security flaws, WEP has been deprecated in favor of WPA

WPA

WPA is the acronym for Wi-Fi Protected Access. It is a security protocol developed by the Wi-Fi Alliance in response to the weaknesses found in WEP. It is used to encrypt data on 802.11 WLANs. It uses higher Initial Values 48 bits instead of the 24 bits that WEP uses. It uses temporal keys to encrypt packets.

WPA Weaknesses

The collision avoidance implementation can be broken

It is vulnerable to denial of service attacks

Pre-shares keys use passphrases. Weak passphrases are susceptible to dictionary attacks.

How to Crack Wireless Networks

WEP cracking

Cracking is the process of exploiting security weaknesses in wireless networks and gaining unauthorized access. WEP cracking refers to exploits on systems that use WEP to implement security controls. There are two types of cracks namely;

Passive cracking– this type of cracking does not affect the network traffic until the WEP security has been cracked. It is difficult to detect.

Active cracking– this type of attack has an increased load effect on network traffic. It is easy to detect compared to passive cracking. It is effective compared to passive cracking.

WEP Cracking Tools

Aircrack– network sniffer and WEP cracker. Can be downloaded from

WEPCrack– this is an open-source program for breaking 802.11 WEP secret keys. It is an implementation of the FMS attack.

Kismet- this can include wireless detector networks, both visible and hidden, sniffer packets, and detect intrusions.

WebDecrypt– this tool uses active dictionary attacks to crack the WEP keys. It has its key generator and implements packet filters.

WPA Cracking

WPA uses a 256 pre-shared key or passphrase for authentications. Short passphrases are vulnerable to dictionary attacks and other attacks that can be used to crack passwords. The following tools can be used to break WPA keys.

CowPatty– this tool is used to crack pre-shared keys (PSK) using a brute force attack.

Cain & Abel– this tool can be used to decode capture files from other sniffing programs such as Wireshark. The capture files may contain WEP or WPA-PSK encoded frames.

General Attack types

Sniffing– this involves intercepting packets as they are transmitting over a network. The captured data can then be decoded using tools such as Cain & Abel.

Man in the Middle (MITM) Attack– this involves eavesdropping on a network and capturing sensitive information.

Denial of Service Attack– the primary intent of this attack is to deny legitimate users network resources. FataJack can be used to perform this type of attack.

Cracking Wireless network WEP/WPA keys

It is possible to break the WEP/WPA keys used to gain access to a wireless network. Doing so requires software and hardware resources and patience. The success of such attacks can also depend on how active and inactive users of the target network are.

We will provide you with essential information that can help you get started. Backtrack can be used to gather information, assess vulnerabilities, and perform exploits, among other things.

Some of the popular tools that backtrack have included;

Metasploit

Wireshark

Aircrack-ng

Nmap

Ophcrack

Cracking wireless network keys requires the patience and resources mentioned above. At a minimum, you will need the following tools

A wireless network adapter with the capability to inject packets (Hardware)

Kali Operating System.

Be within the target network's radius. If the users of the target network are actively using and connecting to it, then your chances of cracking it will be significantly improved.

Patience, cracking the keys may take a bit of sometimes depending on several factors, some of which may be beyond your control. Factors beyond your control include users of the target network using it actively as you sniff data packets.

How to Secure wireless networks

In minimizing wireless network attacks; an organization can adopt the following policies

Changing default passwords that come with the hardware

Enabling the authentication mechanism

Access to the system can restrict by allowing only registered MAC addresses.

Use of strong WEP and WPA-PSK keys, a combination of symbols, number, and characters reduce the chance of the keys been cracking using a dictionary and brute force attacks.

Firewall Software can also help reduce unauthorized access.

Hacking Activity: Crack Wireless Password

In this practical scenario, we are going to use Cain and Abel to decode the stored wireless network passwords in Windows. We will also provide useful information that can be used to crack the WEP and WPA keys of wireless networks.

Decoding Wireless network passwords stored in Windows

Download Cain & Abel from the link provided above.

Open Cain and Abel

Ensure that the Decoders tab is selected then click on Wireless Passwords from the navigation menu on the left-hand side

Click on the button with a plus sign

Assuming you have connected to a secured wireless network before, you will get results similar to the ones shown below.

The decoder will show you the encryption type, SSID, and the password that used.

Summary

Outsiders can see wireless network transmission waves, this possesses many security risks.

WEP is the acronym for Wired Equivalent Privacy. It has security flaws, which make it easier to break compared to other security implementations.

WPA is the acronym for Wi-Fi Protected Access. It has security compared to WEP

Intrusion Detection Systems can help detect unauthorized access

A good security policy can help protect a network.

What is Ethical Hacking EC0-350 Certification Exam?

Ethical Hacking and Countermeasures 6 (CEHv6) Certification is also known as the EC0-350 test. Being conducted by EC Council, this is one of the most prominent certifications in the ethical hacking field.

Security has attained different proportions in today's fast-changing economic scene. The IT department in any organization is one of the most sought after and vulnerable area for hackers. It is the IT which determines the way many organization works, and any breach in its security becomes a big headache. Be it any industry: Banking, Finance, Manufacturing,

Production, Software, Telecommunication, Movies, Entertainment, or Airlines, the security of the data and classified information is becoming famous day by day. Competitors can go to any extent to have the inside knowledge of their opponents.

It is estimated that close to 2 million Ethical hackers were required by 2010-2011 by various industries, to safeguard their assets from dangerous and anti-social elements.

Post 9/11 terrorists attack the US, International Council of Electronic Commerce Consultants or EC Council was established to provide the required cushion against such type attacks on Information Systems around the globe. This formation soon had the support of various subject matter experts around the world, and several certifications and standards were thus developed.

Ethical Hacking and Countermeasures 6 is one among these certifications, targeting the enthusiasts who are aiming for a long term career in this domain. The determination to thwart any hacking intention and the expertise in the programming domain is the only prerequisite for this certification.

There are a total of 150 questions in the EC0-350 test, and a total time of 4 hours is allocated for attempting all the questions. The pass percentage for this exam is 70%.

The successful candidates can go for the Certified Ethical Hacker, Master of Security Science certification.

A Closer Look at Cyber Crooks

I work from my home, the most peaceful workplace I can think of so far. As a Homemaker and part-time Freelance Writer, I submit articles and subscribe to various respectable writing newsletters and do endless research online. But before signing up, I read privacy policies, some brief while others are boringly lengthy.

I am just one among millions of unknown but honest Internet users. Why, to my mind, would I worry so much about anyone in the Internet community making me a target for nasty tricks or harassments?

Nonetheless, I feel safe just knowing that the websites that interest me run a committed sense of policy on security.

But when my computer screen started flashing as if gasping for air and slowly died down months ago, I blamed electric current fluctuations. But when my files disappeared, and my computer turned alarmingly uncooperative, I referred the case to the expert, my husband.

For the first time, I was hit by a computer virus.

It was not as simple as unplugging the computer connection, sleeping on it for a few days while a computer surgeon works on the damage so everything could be good as new.

I saw months of hard work on research, completed manuscripts stashed in my hard disk, long hours spent online, time, money, and effort my spouse has invested in setting up the whole system, all go down the drain and turn into nothingness.

Going through the process of repair and reconstruction is painfully tedious, time-consuming, and costly.

A series of disturbing yet quite interesting cyber intrusions that followed prodded me to quench my curiosity and do personal research on what inspires the behavior behind the waste of skill, time, and resources involved in these damaging cyber pursuits and other electronic petty crimes that make life miserable for honest internet users.

CYBER OFFENDERS

Anyone who enters your home without your consent is committing an offense theoretically. Your computer system is an extension of your boundaries and must not be infringed. When someone gains unauthorized access to your computer in any manner or utilizes computer technology in performing a felony, he/she commits a cybercrime.

The Hacker is always the first person that comes to mind concerning cyberspace violations. After all, who else can be as knowledgeable and bold enough to break into someone else's computer system?

Hackers used to have nobler objectives for their being. In the earlier days of the Computer technology, they were the computer experts/geniuses who tested computer systems, with the owners' consent, for loopholes and recommended better programs or fixed the errors themselves to frustrate any effort to exploit the defective system by more dangerous 'creatures.' They even had the Hackers' Code of Ethics.

There are two types of Hackers: The Ethical Pros, the highly skilled professionals who hire out their skills to organizations concerned about their own network's safety. They represent Hackers of the earlier generation. The other type is the CyberRambos or plain crackers-despised by the Elite Hackers, Crackers crack/brake systems for superficial reasons. (UC San Diego Psycho. Dept.: Computer & Network Resources)

And by whatever name they are called, these cyber felons have become utterly faceless and nameless 'hackers' to their victims.

ON MOTIVES

An online article by David Benton entitled: 'What's Inside a Cracker?' from SANS (SysAdmin, Audit, Network, Security)

Information Security Reading Room, states seven psychological profiles of malicious hackers as documented by Canadian Psychologist Marc Rogers M.A., Graduate Studies, Dept. of Psychology, University of Manitoba and a former Police Computer Crimes Investigator:

Newbie/Tool Kit (NT): new to hacking, have limited computer/programming proficiencies; rely on ready-made pieces of software (tool kits) that are readily available on the Internet;

Internals (IT): disgruntled employees or ex-employees proficient in how the company's internal systems work;

Coders (CD) and Virus Writers: programmers who'd like to see themselves as elite; they write codes but not for personal use. They have their networks to experiment with "zoos." They leave for others to introduce their systems into the "wild" or the Internet. (Hacker Psych 101 by Jeremy Quittner);

Cyber-Punks (CP): antisocial geeks, the most visible, socially inept, and burdened with unresolved anger that they take into cyberspace; they relate better to computers than humans and have better computer skills and some programming capabilities; capable of writing their software, they intentionally engage in malicious acts such as defacing web pages, spamming, credit card number theft, etc.;

Old Guard Hackers (OG): have no criminal intent in its real sense but display an alarming disrespect for personal property with great interest in the intellectual endeavor;.

Professional Criminals (PC) and Cyber Terrorists (CT): most dangerous; they are professional criminals and ex-intelligence operatives who are guns for hire. They specialize in corporate espionage, are exceptionally well trained and have access to state of the art equipment is;

Further, Rogers pointed out that not all Hackers are criminals. He has categorized them as follows: (Jeremy Quittner, Hacker Psych 101);

Old School Hackers: akin to the 1960s style computer programmers from Stanford MIT for whom it is an honor to be a hacker; interested in analyzing systems with no criminal intent; they believe the Internet was designed to be an open system;

Script Kiddies/ Cyber -Punks: wannabe hackers and crackers; use other Cracking programs carelessly with the intent to vandalize and corrupt systems; often caught red-handed because they brag their exploits online.

Professional Criminals: breaking into systems and selling information Is their livelihood; they get hired for espionage; often have ties with organized Criminal groups; not interested in disrupting networks but more on stealing intelligence data;

The list of motives is endless: boredom, illicit thrill, addiction, blackmail or low self-esteem, and a desperate need for recognition from the hacker peer group, all cowardly performed under the protection of anonymity.

"Underlying the psyche of criminal hackers may be a deep sense of inferiority. The mastery of computer technology or the shut down of a major site causing millions of dollars of damage is a real power trip." (J. Quittner, Hacker Psych 101, Hackers: Computer Outlaws)

Jarrold M. Post, a George Washington University Psychiatrist, says: It's (Hackers) a population that takes refuge in computers because of their problems sustaining real-world relationships."

The less information you share on the Internet, the better. But as computer wizards, Hackers will always find ways to reconstruct your identity even with tiny details in their possession.

However, there are varied ways by which you, a legitimate Internet user, can be protected. Know the warning signs and get educated on how to thwart any attempt to victimize you. Don't take the wired blows sitting down.

"Constant awareness and updating of knowledge is the best defense to any attack," wrote Shayne Gregg, CA (NZ), CISA,

CMC, in 'A Response to Recent Cyber Attacks.' (Information Systems Audit & Control Association InfoBytes)

I recommend The Complete Idiot's Guide to Protect Yourself Online by Preston Gralla, Executive Editor, ZDNet. It is comprehensive, easy to understand, and a must for every Internet user's library.

HACKERS, CYBERPUNKS, et al

Hackers or crackers do not monopolize cyber Crime. The pedophile, thief, or drug dealer in your community who hire computer experts to carry out their illegal activities online are as guilty and despicable.

Just like the criminals roaming out in the real world, Cyber felons are a bunch of psychologically imbalanced and misguided citizens who happen to have the dexterity to commit electronic transgressions or hire a computer expert to do the job and will never get enough despite their Cyber Glory and 'conquests.'

Still, the tendency to commit a crime lies hidden in wholesome images, while the unsuspecting are often caught by surprise. What you don't see is sometimes what you get.

Hackers cannot be strictly stereotyped. Peter Shipley, Chief Security architect for the Big Five firm KPMG, avers: "I know many hackers, including one who spends an hour and a half in the gym every day. He is built. I know of knock-down gorgeous women who are hackers."

No Exceptions

Whenever great tech-crazed folks travel through your wires to make your computer system malfunction, steal your identity, or get paid to give you trouble, it's a sign that you do not take the needed precaution whenever you log in.

When I asked myself quietly back then, "Why me?" I guess the reply would be, "And why not?" As with most inventions, the Internet is being abused and mishandled. And as always, a helpless victim completes the drama.

Anyone can be a casualty at random regardless if one is honest, educated, high profile, residing at the far end of the globe or a Stay-at-Home- Mom working hard and peacefully from her abode.

Hackers won't care how his/her prey will feel.

But I am still hoping that such an impressive brainwork will be put to good use by present-day hackers, just how their predecessors intended Hacking to be used.

The Right Path For A First Time Linux User

Are you new to Linux or thinking of using it for the first time? Hold on! What the heck I'm saying here! There's no word called " New to Linux" or " first time Linux user." Without your conscious, you probably use it every single day! And you thought Linux meant for the programming nerds, hackers and going through Linux says using that suitable old green terminal!

That's racist, you know.

The main question should be- are you new to personal Linux computing? Well, if your answer is "Yes," then worry not, a superior operating system is ready to be served for his only master. See what I did there? If not, I mean to say that you and only YOU are the owners of your hardware and software. No one was going to install some crappy app that you don't need or change the system setting while you are enjoying good old " funny kitty video" on the internet!

In the world of Linux personal computing, there are a plethora of choices to choose from. People from Linux planet call this "Distributions." What this means is, while the primary system 'Kernel" is identical, the look & feel and the entire ecosystem can be different.

My personal favorite, for my day-to-day desktop work at least, is Kali Linux. It is a Debian-derived Linux distribution designed for digital forensics and penetration testing. However, for my personal use, I prefer Linux Mint or Elementary OS. But here are some others you may have heard of:

Fedora

Zorin

openSUSE

Debian

Those are the most significant distribution in terms of users. However, as a beginner, you should use " Linux Mint cinnamon edition." It closely resembles your windows PC, and if you are coming from the world of fruit, I will suggest giving a try to the "Elementary OS" or "Deepin OS." They closely resemble the Mac ecosystem.

Stick with Linux Mint: From the first day of migration, it is recommended that you stick with distributions like Linux Mint, Zorin, Linux Lite e.t.c. There are easy to install and use, and they have a massive number of the online user base. These users are relatively knowledgeable and kind, ask them what problem you are facing, and you will have the correct answer within hours if not within minutes! Linux Mint comes with reasonably decent software out of the box. These include libre office (a free & open-source office suite), Thunderbird (Email client), Rhythm (Music Player), and Firefox(you can easily install chrome and chromium). As you get too familiar with the Linux environment, you might end up experimenting with different distros and DE (Desktop Environment). However, for now, it is a good idea to stick with Linux Mint and slowly understanding how Linux works.

Immerse Yourself: the Best way to set a relationship with Linux is to make it your daily driver. Without any doubt, the first few day's rides would be bumpy and strange, so is everything new and beyond one's comfort zone. A distribution like Linux Mint, Zorin & Ubuntu try to make the journey from Windows or Mac into the universe of Linux smooth and magical! Pretty soon, I can assure you that you will be wondering why you ever used anything other than Linux!

Don't be scared of the terminal: Distributions like Ubuntu and Linux Mint are made so that you never really have to open the terminal command line if you don't want to. However, getting to know the command line is profoundly encouraged, and it's not nearly as painful as it looks at first. The command line is better and more productive than the Graphical User Interface (GUI) in many cases. What takes several clicks, scrolls, keystrokes, and more clicks in the GUI can usually be accomplished with a single terminal command. That's the simplicity!

Ally, with Google: With the passing time, you will come across something in Linux that you desire to do; however, you aren't sure what method you should follow. This is where Google will become your best buddy! If there's something you can't figure out how to do in Linux, someone other than you has run into that same problem before. The official Ubuntu Wiki and AskUbuntu forums will be controlling your search outcomes. Conveniently, Linux Mint is built on Ubuntu, so whatever solution works in Ubuntu is virtually guaranteed to work in Linux Mint as well.

There's a lot more I want to say; however, I think you will learn them eventually. In conclusion, I would like to state that follow Linux blog pages, follow their social media. Know about themes and icon packs and always desire to do something new and creative. Have a happy journey to the wonderland. Thanks for reading!

CHAPTER 7

PRACTICAL HACKING

An Overview of Ethical Hacking

Does the word hacking scare you? Ironically it is hacking but legal hacking that is doing us right. If this is your first article on piracy, then surely you will get some potential insight on hacking after reading this. My article gives a simple overview of ethical hackers.

The term ethical hacker came into the surface in the late 1970s when the government of the United States of America hired groups of experts called 'red teams' to hack its hardware and software system. Hackers are cybercriminals or online computer criminals that practice illegal hacking. They penetrate the security system of a computer network to fetch or extract information.

Technology and the internet facilitated the birth and growth of network evils like viruses, anti-virus, hacking, and ethical hacking. Hacking is a practice of modification of a computer hardware and software system. Illegal breaking of a computer system is a criminal offense. Recently a spurt in the hacking of computer systems has opened up several courses on ethical hacking.

A 'white hat' hacker is a moral hacker who runs penetration testing and intrusion testing. Ethical hacking is legally hacking a computer system and penetrating its database. It aims to secure the loopholes and breaches in the cyber-security system of a company. Legal hacking experts are usually Certified Ethical Hackers who are hired to prevent any potential threat to the computer security system or network. Courses for ethical hacking have become widely popular, and many are taking it up as a solemn profession. Ethical hacking courses have gathered huge responses all over the world.

The moral hacking experts run several programs to secure the network systems of companies.

A moral hacker has legal permission to breach the software system or the database of a company. The company that allows a probe into its security system must give legal consent to the right hacking school in writing.

Moral hackers only look into the security issues of the company and aim to secure the breaches in the system.

The school of moral hackers runs vulnerability assessment to mend loopholes in the internal computer network. They also run software security programs as a preventive measure against illegal hacking.

Legal hacking experts detect security weakness in a system which facilitates the entry for online cybercriminals. They conduct these tests mainly to check if the hardware and software programs are sufficient enough to prevent any unauthorized entry.

The moral experts conduct this test by replicating a cyber attack on the network to understand how strong it is against any network intrusion.

The vulnerability test must be done regularly or annually. The company must keep a comprehensive record of the findings and check for further reference in the future.

Examples of Ethical Hacking - How Hacking Can Improve Our Lives

If you are looking for examples of ethical hacking, then read on!

It's funny because the concept of carrying out what is a malicious attack ethically has certainly evolved people's understanding on the subject of hacking. People tend to immediately associate this with negative actions and intentions because they only know the adverse effects. In short, most will believe there can be little or no positive application for it, but of course, that is not true.

When used for good, it's excellent!

When used as a means to improve an individual or a company's online defenses, we find this "malicious act" rather beneficial. The practice of breaking into, or bypassing an online system or network to expose its flaws for further improvement is entirely ethical (and you can make a beautiful living doing it too.)

Examples of ethical hacking include exploiting or exposing a website to discover their weak points. Then report your findings and let the appropriate person fix those vulnerabilities. Later in the future, should they come under attack, they will be that bit safer. You are preparing them for any real threat of attack because you are eliminating the areas which could be exploited against them.

There are many examples of ethical hacking, including one which happened in the early days of computers. Back then, the United States Air Force used it to conduct a security evaluation of an operating system. In doing so, they were able to discover flaws like vulnerable hardware, software, and procedural security. They determined that even with a relatively low level of effort, their protection can by-passed, and the intruder would get away with valuable information. Thanks to ethical hacking, they were able to stop such an incident from happening. The people who carried out this task treated the situation as if they were the enemy, doing all they could to break into the system. This way, they could determine exactly how secure their system was. This is perhaps one of the best examples of ethical hacking because they were sanctioned by the people who were responsible for the creation of the said online system. They recognized the need for such action because they know that there are a lot of people capable of doing the same thing or inflicting the same harm to their system.

From all the examples of ethical hacking, perhaps you can relate to the practices of known Operating Systems being used today. Makers of these Operating Systems perform their ethical hacks to their systems before actually launching their products to the public. This is to prevent possible attacks that could be perpetrated by hackers. This is somehow a means of quality control during the system's development phase, to make sure that all the weaknesses of their Operating Systems are covered since it will be marketed for public use. Ethical hacking is an advantageous approach in defending your precious online systems. By tapping into the abilities and potential of white hat hackers, you can take on and prevent damages caused by the real hackers.

Why Should We Consider Ethical Hacking Seriously?

While talking about hacking, what do we tend to imagine? A silhouetted figure in hoodie typing something in the computer, a black screen, innumerable codes, a dark indoor, right? In movies, it just takes a few seconds to breach into a system and get all the data. But, in reality, it takes lots of sweat and blood to carry out the procedure called 'Hacking.'

It takes immense hard work, skills, knowledge, and passion for becoming a professional Ethical Hacker. Now, the question arrives, how can interfering with someone else's database be ethical? Though it sounds like an oxymoron, the world indeed needs white hat hackers now more than any time before.

Business houses, law enforcement cells, Government houses require skilled professional, ethical hackers.

With the advancement of technology, like IT outsourcing, cloud computing, virtualization, we are exposed to various security threats every day. In that case, the networking experts are hired to protect the database of a particular organization from potential harmful exploiters. Data exploitation can lead to more significant damage to reputation and financial loss for any company. Now ethical hacking is one of the most popular security practices performed regularly.

Cybercrimes have increased massively in the last few years. Ransomware like WannaCry, Petya is making news every day with their other variants, and it will not be an exaggeration to say that they are here to stay increasing their muscle power to cause more harm. Phishing schemes, malware, cyber espionage, IP spoofing, etc. are prevalent now. To safeguard data, companies need to adopt a proactive stance.

With the ever-increasing popularity of cloud comes baggage of security threats. Now, when business organizations are using cloud services like Google Drive, Microsoft Azure, or Dropbox, they are storing sensitive data on a third-party tool, which may or may not work in their best interest. Using third file-sharing services allows the data taken outside of the company's IT environment. This often leads to several security threats, including losing control over sensitive data, snooping, key management, data leakage, etc.

Almost every one of us is active on various social networking sites. We actively share our whereabouts, interests, address, phone numbers, dates of birth there, and with the information, it is easy for cybercriminals to figure out the victim's identity or steal their passwords. A study reveals, around 60,000 Facebook profiles get compromised every day. Social media users are likely to click on anonymous links shared by friends or someone they trust. This is an old method of exploiting the victim's computer. Creating fake Facebook 'like' buttons to webpages is also a viral method of cybercrimes.

The definition of network forensics and ethical hackers has been evolved over time. Many organizations are yet to realize that the cost to protect the company database is much less than dealing with a grave cyber attack to recover all data. Prevention is always better than cure. Network forensics and ethical hackers are hired in IT sectors to continuously monitor and identify potential vulnerabilities and take action according to that.

Six (6) Sure Signs You Have Been Hacked

There are several ways in which antivirus scanners try to detect malware. Signature-based detection is the most common method.

This involves searching the contents of a computer's programs for patterns of code that match known viruses. The anti-virus software does this by checking systems against tables that contain the characteristics of known viruses. These tables are called dictionaries of virus signatures.

Because thousands of new viruses are being created every day, the tables of virus signatures have to be continuously updated if the anti-virus software is to be effective. But even if the software is being updated daily, it usually fails to recognize new threats that are less than 24 hours old.

To overcome this limitation and find malware that has not yet been recognized, anti-virus software monitors the behavior of programs, looking for abnormal behavior. This technique is called heuristics. The software may also use system monitoring, network traffic detection, and virtualized environments to improve their chances of finding new viruses.

Nevertheless, anti-virus software is never 100 percent successful, and every day new malware infects computers throughout the world.

Getting hacked

There are three main ways you can get infected with malware.

These are: (a) running unpatched software, i.e. software that you have failed to update; (b) falling for a desirable freebee and downloading a Trojan horse along with the freebie; (c) responding to fake phishing emails.

If you can manage to avoid these three failings, you won't have to rely so much on your anti-virus software.

Expecting that someday someone will release anti-virus software that can detect all viruses and other malware with complete accuracy is a vain hope. The best you can do is to keep your security up to date, avoid the three main ways you can get infected, and learn to recognize the signs that suggest your computer has been hacked so that you can take appropriate action.

Here are some sure signs you've been hacked and what you can do about it.

[1] Fake antivirus messages

A fake virus warning message popping up on the screen is a pretty sure sign that your computer has been hacked-provided; you know it's fake. (To be able to recognize a phony warning, you need to know what a certain virus warning from your anti-virus software looks like.) The signal will reassure you by saying that it is can scan your system to detect the malware.

Clicking no or cancel to stop the scan won't help because your computer has already been compromised. The purpose of the fake virus warning (which will always find lots of illnesses that need to be eliminated) is to lure you into buying their virus removal service or other product.

Once you click on the link provided for that purpose, you will likely land on a very professional-looking website. There you'll be invited to buy and download the product by filling in your credit card details.

Bingo! As well as having complete control of your system, the hacker now has your personal financial information.

What to do: as soon as you see the fake virus warning message, turn off your computer. Reboot it in safe mode (no networking) and try to uninstall the newly installed software (which can often be uninstalled just like a regular program).

Then, whether you succeed in uninstalling the rogue program or not, restore your system to the state it was in before you got hacked. Nowadays, you can typically revert to a previous state with just a few clicks.

Once you have turned back the clock, to speak, healthily restart your computer and make sure that the fake virus warning has gone. Then do a complete anti-virus scan to eliminate any traces of the malware.

[2] Unwanted browser toolbars

Finding your browser has a new toolbar is probably the second most common sign of being hacked. Unless you recognize the toolbar and know that you knowingly downloaded it, you should dump it.

Very often, these toolbars come bundled with other software you download. Before you begin a download, you should always read the licensing agreement, which may contain a clause allowing other software to be downloaded with the software you want. Hackers know that people seldom read these agreements, yet having these kinds of requirements makes the downloading entirely legitimate.

What to do: Most browsers allow you to remove toolbars. Check all your toolbars and if you have any doubts about a toolbar, remove it. If you can't find the bogus toolbar in the toolbar list, check to see if your browser has an option to reset it back to its default settings.

If this doesn't work, restore your system to the state it was in before you noticed the new toolbar, as described in the previous section.

You can usually avoid malicious toolbars by making sure that all your software is fully up-to-date and by being ultra-cautious when you are offered free software for downloading.

[3] Passwords changed inexplicably

If you discover that a password you use online has been changed without your knowledge, then it is highly likely you have been hacked. If not, your internet service provider (ISP) has been compromised.

Above all, you need to amend your behavior for the future. Reputable websites will never ask for your log-in details by email. If they do appear to do so, do not click on the link in the email. Instead, go directly to the website and log on using your usual method. You should also report the phishing email to the service by telephone or email.

[4] Unexpectedly finding the newly installed software

If you find new software on your computer that you don't remember installing, you can be reasonably sure that your system has hack.

Most malware programs nowadays are trojans and worms which install themselves like legitimate programs, usually as part of a bundle with other applications that you download and install. To avoid this, you need to read the license agreement of the software that you do want to install strictly to see if it comes with 'additional' software.

Sometimes you can opt-out of these 'free' extras. If you can't, your only option, if you want to be sure you are not going to be hacked, is not to download the software you do want to install.

What to do: the first thing you should do (in Windows) is to go to Add or Remove Programs in the Control Panel. However, the software program may not show up there on the list. In so, there are plenty of applications available on the Internet (usually for free), which will show all the programs installed on your computer and enable you to disable them selectively.

This approach has two problems. Firstly, these free programs cannot guarantee to find every installed application. Secondly, unless you are an expert, you will find it hard to determine what are and what are not legitimate programs.

You could, of course, disable a program you don't recognize and restart your computer. If some functionality you need is no longer working, you can re-enable the program.

However, your best bet, in my view, is to stop taking risks (and wasting time) by calling an expert technician at an online computer maintenance company who can scrutinize your system for illegitimate programs and delete them as necessary.

[5] Cursor moving around and starting programs

Cursors can walk around randomly at times without doing anything in particular. It is usually due to problems with hardware.

But if your cursor begins moving and makes the correct choices to run particular programs, you can bet your last dollar that you've been hacked and that humans are controlling your mouse.

The hackers who can take control of your computer in this way can start working in your system at any time. However, they will usually wait until it has been idle for a long time (e.g., during the early hours of the morning) before they start using it, which is why it is vital that you turn off your computer at night and disconnect it physically from the internet.

Hackers will use their ability to open and close programs remotely to break into your bank accounts and transfer money, buy and sell your stocks and shares, and do all sorts of other nefarious deeds to deprive you of your treasure.

What to do: If your computer suddenly swings into action some night, you should turn it off as soon as possible. However, before you try to find out what the hacker is interested in and what they are trying to do. If you have a digital camera handy or a smartphone, take a few pictures of the screen to document what the hacker is doing.

After you have closed it down, disconnect your computer from the internet and call for professional help. To solve this problem, you will need expert assistance from an online computer maintenance firm.

But before you call for help, use another computer that is known to be good, to change all your log-in details for your online accounts. Check your bank accounts, stockbroker accounts. If you discover that you have lost money or other valuables, call the police and make a complaint.

You have to take this kind of attack seriously, and the only option you should choose for recovery if a complete clean-out and re-installation of your operating system and applications.

But before you do so, if you have suffered financial losses, give a forensic team access to your computer so they can check what took place. You may need a report from them to recover your monetary damages from your insurer, banker, broker, or online merchant.

[6] Anti-virus program, Task Manager or Registry Editor disabled and won't restart

Stuff can happen, so one of these three applications could go wrong on its own. Two of them might go wrong at the same time in a million-to-one coincidence. But when all three go wrong together.

Much malware does try to protect itself by degrading these three applications so either they won't start or they start in a reduced state.

What to do: you cannot know what happened, so you should perform a complete restoration of your computer system.